Falling for Icarus

By the Same Author

Stalin's Nose
The Oatmeal Ark
Under the Dragon
Next Exit Magic Kingdom

Falling for Icarus

A Journey Among the Cretans

RORY MACLEAN

VIKING
an imprint of
PENGUIN BOOKS

VIKING

Published by the Penguin Group
Penguin Books Ltd, 80 Strand, London WC2R 0RL, England
Penguin Group (USA) Inc., 375 Hudson Street, New York, New York 10014, USA
Penguin Books Australia Ltd, 250 Camberwell Road,
Camberwell, Victoria 3124, Australia
Penguin Books Canada Ltd, 10 Alcorn Avenue, Toronto, Ontario, Canada M4V 3B2
Penguin Books India (P) Ltd, 11 Community Centre,
Panchsheel Park, New Delhi – 110 017, India
Penguin Books (NZ) Ltd, Cnr Rosedale and Airborne Roads,
Albany, Auckland, New Zealand
Penguin Books (South Africa) (Pty) Ltd, 24 Sturdee Avenue,
Rosebank 2196, South Africa

Penguin Books Ltd, Registered Offices: 80 Strand, London WC2R 0RL, England

www.penguin.com

First published 2004
1

Set in 12/14.75pt Monotype Bembo
Typeset by Rowland Phototypesetting Ltd, Bury St Edmunds, Suffolk
Printed in Great Britain by Clays Ltd, St Ives plc

A CIP catalogue record for this book is available from the British Library

ISBN 0-670-91483-5

to KDM and FCM

Come to the edge, he said
They said: We are afraid
Come to the edge, he said
They came
He pushed them . . . and they flew.

Guillaume Apollinaire

RODOPOU PENINSULA
AKROTIRI
SOUDA BAY
Kissamos
Kolimbar
Maleme
Gerani
Galatas
Hania
Souda
Kalives
Almirida
APOKORONAS
DRAPANOKEFALA
FALASSARNA
Kefali
Fournes
Gavalohori
Vamos
Vrysses
Kefalas
Anissari
Exopoli
Rethimnon
Skaleta
Perama
Zoniana
Georgioupoli
Pasalites
PSILORITIS
LEFKA ORI
(WHITE MOUNTAINS)
Enneachora
SFAKIA
Paleohora
SAMARIA GORGE
Aradena
Horasfakion
Vouvas
FRANGOKASTELLO
AMARI VALLEY
Amari
MT. IDA
Gerakari
Ano Meros
NIDA PLATEAU
Timbaki
Matala
MESSARI PLAIN
LIBYAN SEA

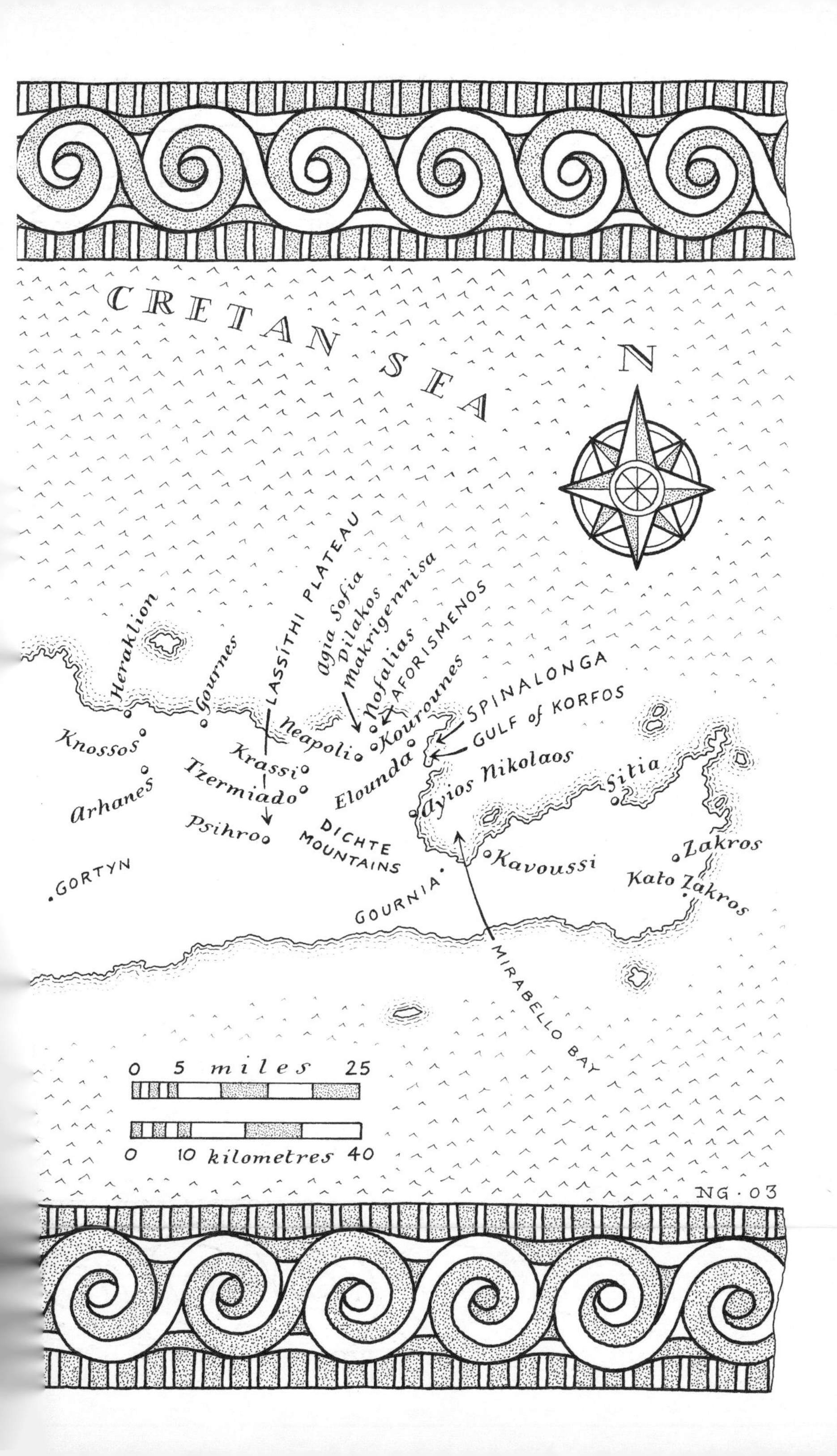
CRETAN SEA
N
Heraklion
Gournes
LASSITHI PLATEAU
Agia Sofia
Dilakos
Makrigennisa
Nofalias
AFORISMENOS
Kourounes
SPINALONGA
GULF of KORFOS
Knossos
Neapoli
Krassi
Tzermiado
Elounda
Ayios Nikolaos
Sitia
Arhanes
Psihro
DICHTE MOUNTAINS
Kavoussi
Zakros
Kato Zakros
GORTYN
GOURNIA
MIRABELLO BAY
0 5 miles 25
0 10 kilometres 40
NG · 03

1. Without Wings

This is what happened. It was after three and I couldn't sleep. Another broken, black night. I went for a walk in the ruins. A raw breeze blew up and I decided to climb the rocks. I found a foothold, the mark of an ancient chisel, and pulled myself up the old wall.

In the dark I moved by touch, caught hold of a root, searched for a second step. I felt rather than saw the hand-cut limestones. The sheer blocks were cold against my cheek. My hand reached for a crevice, dug into it and lifted the weight of body and heart. A speck of mortar blew into my eye. I worked up towards the glow of sky. I reached an arm over the parapet and clambered to my feet.

I stood at the edge suspended between heaven and earth. At my feet spread the silhouettes of islands, mulberry shadows in a moonlit sea. Behind me a twisted cord of cloud tethered the peaks, separating them from the foothills, dividing silver snow from green olive groves. The lower flanks of the mountains were flecked with white light – villages not yet asleep. Higher up the slope the solitary warm pinpoint of a shepherd's fire flickered and died. Above it the sky was crowded with stars, a thousand sparks piercing the night.

I turned into the quickening breeze. I took a deep breath. I balanced on my heels and tucked my toes over the precipice. Eyes shut. Head up. Arms out. A sudden gust pushed me back and I leant into it.

I wanted to feel the lift. I was ready for the lightness. The soles of my feet tingled. Legs together. Palms angled into the wind. My stomach tightened in anticipation of the rush of flight. I laughed for the first time in months.

'Got you!'

'I'm falling.'

Eyes wide open. I spun round, scrabbling to catch the wall. A stranger's hands gripped my ankles. Horny, earthy hands which had knocked me off balance.

'Hang on,' shouted the voice above the wind.

'What the hell are you doing?' I managed to say. I'd thought I was alone.

'Saving your life, you bastard.'

This is how I met Yióryio.

I wasn't trying to kill myself. Really. I was groping for a way forward. It had been six months since my mother's death. A year since her cancer had been diagnosed. My wife and I had taken her into our house. We had nursed her, bathed her and held her hand as she passed away in the pale green English bedroom. The loss numbed me. It splintered my confidence and crippled my imagination. But her death had one surprise in store for me. A sudden, starved passion that no one in their right mind could have predicted.

When my mother had taken her last breath, and the swallows swept out from their nests under the eaves, I wanted to fly. From that moment I needed to feel white wings lift me into a warm spring sky. The compulsion was the single clear certainty in my now dislocated life.

Which is how there came to be a stranger gripping my ankles, dangling me over a cliff face, in the chaotic ruins.

'You can't jump,' said Yióryio, a truss of chicken lashed to his belt. 'You don't have wings.'

'Not yet,' I yelled back at him, as he pinned me down on the cold, black wall.

'Excuse me,' he said, tightening his grip, 'but I think you are a little crazy.'

2. Beyond My Control

'The Crete is best place for you,' Yióryio bellowed, making himself heard above the din, delighted for all the world. In his *kafeneion* a dozen men surrounded me, all shouting at once, none listening to the other. 'The very best place in world to build an aeroplane.'

It was the first of many tall stories.

'Because of Daedalus and Icarus?' I asked as he refilled my glass. I knew it would be an insult to refuse his fiery *tsikoudiá*, even before breakfast.

'Because I am here.'

An hour earlier on the ancient walls Yióryio had released my ankles and taken me in hand. I had told him about my plans and he had offered to help. In fact, he had pulled at his wild moustache and insisted on helping, inviting me to his village, promising to rent me a house, volunteering his uncle's workshop and cousin's truck. His enthusiasm dismayed me. If he questioned the practicality of my undertaking there was no doubting his willingness to assist me in achieving it. So I had followed him off the mountain and down through the wind-blown almonds, ahead of the rain, and joined the early morning scrum of villagers at the café.

'The *pilótos* and I will be neighbours,' declared Polystelios, or Stelios-of-Many-Talents, reaching out to shake my hand for the third time. He was a failed beekeeper who'd earned his nickname for being a master of nothing. A woolly hat capped his hollow, buffed plaster face. 'You have flown many times, of course,' he asked me.

'Only in economy.'

'Then you've built a flying machine before?' said Socrates the shepherd, taciturn and hard-muscled, fingering his amber *kolmbolói* worry beads.

'I've made a bookcase.'

'Did it fly?' jibed 'Little' Iánnis, the sullen, yellow-eyed, one-armed barber, looking up from his cards.

Kóstas the village policeman stole a look at Iánnis' hand and asked, 'But you have a licence?'

I shook my head.

'In the Crete laws is very elastic,' interrupted Yióryio with a happy grin of anticipation. He seemed unconcerned by legal niceties. Or why I had to fly. 'No man ever say that he cannot do something. We say only that we can. This is Cretan way.'

'Maybe I am mad,' I admitted, propelled by intuition, by emotion, by myth, 'but in my head I think of only one choice.'

'I know the *pilótos* will fly,' insisted Polystelios, raising a glass to me, 'as an eagle above crows. *Yiá sta khéria sas.*' Meaning 'Health to your hands.' May your talent flourish.

'And only one time,' I added, my lunatic Greek entangled with their fitful English. 'Any bigger and I am pushing my luck.'

Yióryio stood with his feet planted on the ground, on the

unrendered concrete floor, as if he was growing out of it. He was a sturdy man, arrogant with audacious mahogany brown eyes and cheeks which turned into beef tomatoes when he laughed. His hair was long for a Greek. It might not have been cut for years. Tight, Victorian ringlets – shot through with silver grey – covered his short, stubborn neck. A gold cross nestled against the black thatch on his chest.

Next to him were ill-fated Polystelios, whose jerky movements brought to mind an arthritic marionette, and ample Papá Nikos, the bearded village priest, dreaming not of the riches of heaven but of his Alpha Bank high-interest deposit account. In the coming months I never once saw him offer to buy a drink. Socrates the shepherd sat between them and barren Iánnis, Yióryio's barber cousin with sharp nose and wit, home from Athens.

'Before the war someone else wanted to be an Icarus,' said Papoos, tapping his cane on the floor to attract attention. The back-bent grandfather wore high black boots and a white beard clipped beneath his chin like a Minoan.

'A Cretan?' I asked, needing to believe him.

'He made wings and climbed to the top of Hania lighthouse then became frightened and wouldn't jump. So his neighbours pushed him.'

'Did he survive?'

'He fell into the water,' reported Papoos as the men rolled about with laughter. 'It made him a little crippled, but if you can fly it doesn't matter if you have one or two legs.'

Man had first flown in this brightening sky, out of the twilight where history and legend met. The story was a time-worn link in the chain which had brought me to Crete. Daedalus and Icarus had taken to the air on wings made of feathers and wax. Icarus had flown too close to the sun, the wax had melted and he fell into the sea. Daedalus, his father –

the great inventor whose works, Plato wrote, were touched with divinity – had flapped on across the Aegean to a new life.

'*Kalós orísate*,' Yióryio cheered in welcome, cocking his head as he refreshed the hospitable glasses of *tsikoudiá*.

'*Kalós sas vríkame*,' I replied. It is good that I have found you.

These men were the players in my unfolding drama, their voices clear and animated, their stories of uncertain authority, their alcohol mightily potent. Their weathered faces reflected the island's landscape: rugged, austere, defiant. Crete was at once foreign and familiar; a confident, alien society against which my confusion might be cast into relief. Like a comical pantheon the villagers were to guide me in my aerial quest. They would also help me to accept that nothing lasts forever. Except perhaps Cretan tales.

Under tin-framed photographs of stern-faced ancestors and a pair of Dionysian bunch-o'-grapes wall lights Yióryio passed round plates of roasted pumpkin seeds and wild capers pickled in vinegar.

'All of Europe is the Greece. All of Greece is the Crete. All of Crete is Anissari,' he crowed, naming their village. 'So Anissari is the centre of Europe.'

Yióryio's laugh gusted out of him, released like Aeolus' winds. His repartee hovered on the wrong side of impertinence. He enjoyed the sound of his voice, holding sway over proceedings, indulging his jaunty, dubious sense of humour, making us willing hostages of his hospitality. I tried to withdraw my glass but he gripped my arm and refilled it.

'What do you need for this flying machine?' asked Socrates, his reserve eased by the drink, flicking his beads with increasing excitement.

'Only bits and pieces,' I said. I intended finding a simple design which could be made by hand with basic tools. 'Some wood and fabric, some wire, a few nuts and bolts.'

The men wanted to know the likely wingspan, engine size, weight and range. My tentative answers seemed to raise no concerns, at least no one suggested that there might be any problems.

'I also need a big space to build it and, of course, somewhere to take off. Where can I find one hundred metres of straight road?'

'You being in the Crete before?' asked Yióryio, not answering my question.

'My wife and I arrived last night.'

'Then you not know that Crete is most beautiful continent in world.'

'Continent?'

'You not yet see our jungles and deserts, our fifteen mountain ranges and one hundred gorges. You not know our lost civilizations or pig hippopotamus, long time dead. All these wonders make us a continent, the smallest and greatest continent of world,' he assured me without a wink. 'This, and everything I will ever tell you, is true.'

His extravagance animated the men who now began banging the tables in hilarity. Papoos called for more drink. Polystelios mounted a rush-bottomed chair and began to shout. At the same moment a roll of thunder surged up the coast, encircling the whitewashed village and shaking off the electricity. The winter rain beat snare drum rolls across the roof. I had never heard so much noise at such an early hour. There was nothing to dampen it in the *kafeneion*; no wood, no curtains, no modesty. And it was, of course, great sport to get me legless.

'I'm sure he will fall down,' hissed Papoos in Greek.

'What do you care, old man?' asked Iánnis.

'As long as he doesn't crash on my vines.'

'*Ti na kanoume?*' shrugged Yióryio. What can we do? 'It's

not our problem if he kills himself. He has a vision and we must help him achieve it.'

'Will his vision be on television?' asked Polystelios.

Yióryio winked and stroked his moustache.

'I'm sorry,' I interrupted, 'but my Greek is poor.'

'No problem. We all speak good Greek.' He went on, 'In Crete every man need *koumbáros* – a best man – to look for him and his family. I will be your *koumbáros*.'

'Even though you think I'm crazy?'

'Because I *know* you are crazy.'

A flock of sheep broke through a gate of cut gorse, spilled past the window and surrounded Kóstas the policeman's bumblebee-yellow bulldozer. Polystelios threw up his hands and limped out into the storm.

In the ruckus Iánnis, who had guzzled his drink, leant towards me and snarled, 'Why didn't you take the brain tablets before coming to Crete?'

'I have flown many times in my life,' volunteered the priest, 'thanks be to God.'

'And I went to Athens once,' countered Papoos, pointing his cane towards the ceiling and hitting the fan. 'I didn't like it at all.'

As Polystelios herded his sodden animals back into a broken field, a slender figure stooped out of the weather and into the *kafeneion*, hoisting his trouser legs from the knee to expose bone-white ankles. Ulysses the village idiot perched on a stool, knotted his feet around a rung and stared into space between us. His features were sensuous: full lips, drooping green eyes, long bovine eyelashes. He held to his ear a silent transistor radio, which was either broken or had no batteries. Yióryio fetched him a foil-wrapped chocolate pastry and his face lit up like a child's.

'Now where you will live?' Yióryio asked me.

'We've rented a room in Gavalohori.' The next village.

'*A pa pa pa*,' he said without approval. 'I don't like people who brag about their own place but it is sure that ours is best village in Crete.'

The other men roared in considered agreement.

'There are many places here in Anissari where you can build your aeroplane and behind my *kafeneion* is a free house.'

'Free?'

'Not expensive at all. You and your wife can stay a week, a month, as long as always. You are our guest.'

It seemed an ideal place for me to start to rebuild myself; a subsistent village surrounded by white mountains near to the sea in an ancient corner of Crete. Here there were curious helpers, a carpenter, small agricultural engines and Yióryio. I started to thank him but he raised a hand to stop me.

'In my *kafeneion* you talking to everyone,' he declared, embracing the circle of men in his broad arms, 'whatever their language. All of us we help you make your machine. Some bring wood. Some bring metal.'

'I will bring you a grave,' hissed Iánnis.

'Then one day you will fly from outside this door,' said Yióryio, 'and all the Crete sees our village on television.'

'And you will come with me?' I asked him as he reached again for the bottle.

'Me?' He shook his head. 'You fly. I go on Rosetta.'

'Rosetta?'

'My donkey.'

And so I fell to earth in Anissari.

3. Messa Anissari

At midmorning when I stumbled out of the *kafeneion* I was already flying. I soared beneath the mackerel sky, down from the ancient ruins, over the bone-bare plateau of scorched stone tinted violet by spring iris. The storm clouds parted and a sweeping bay uncurled beneath me, its slate-coloured waters clipped into peaks by the wind. The sun broke cover to catch the wave tops in a thousand flashes of silver.

With my head spinning I sailed over the flat earthen roofs of Anissari, between water tanks and clothes lines, above whitewashed walls and black-scarved *yiayiáthes*, wrinkled as walnuts, peeling potatoes by faded blue doors. The sound of sawing lifted up from Manólis' workshop. Iánnis threw a stone at his chained dog. A pair of warblers snatched insects in mid-flight, turning on the air, clicking their beaks with each catch, nesting in a hole in the cracked chapel ceiling. I glided past its squat bell-tower, hovered above a lush plain of vines and a hidden valley of cypress – then walked into a telephone pole and landed flat on my face.

Anissari is a huddled, intimate hamlet tucked away in a fold of hills, forgotten beneath the snow-capped peaks of the Lefka Ori, the majestic, lunar White Mountains. Beneath them the villagers doze away the winter in front of gas fires, wander about in ill-fitting caps, try to keep warm by thinking of summer. A passing motorist might blink and miss Anissari altogether, later claiming that the village didn't exist at all. But its insignificance was lost on the inhabitants themselves for whom, as I'd heard, it was the centre of Europe.

In fact, Anissari is three distinct villages which glare at each other across the rim of a shallow, leafy basin. The largest is Pano – or Upper – Anissari. It calls itself 'Little Paris' and looks west towards Hania, the old capital of Crete. The mushroom cupola of its Agios Ioannis has stood for a century as the tallest building in the valley. Once wealthy and proud, Pano Anissari considers itself to be a cut above the rest, even though its fortunes have faded and its young people have moved away to Heraklion. Its school had closed, the low lazy arches of the olive oil factory had crumbled into the caper bushes and the big house had been inherited by an Athenian widow who never came to visit. Only a few dozen ageing villagers remained behind the high walls, haunting the narrow lanes, tending pots of hibiscus and the graves of their beloved dead.

A stone's throw to the east is Kato Anissari, known to locals as 'Anissari of the Day' because of the dozen children who live in the village. Plump black hens and fretful guinea fowl picked through its verdant gardens. Dogs barked at its gates. Two bread vans called by every day of the year, except for Good Friday. The yellow school bus stopped on its way to

Vamos. Unlike 'Little Paris', Lower Anissari turned its face towards the valley, and the olives, vines and sheep which sustained its hundred families. Its grapes produce the valley's finest wine, or so its mayor assured me. Its goats' milk was said to be favoured by the cheese makers of nearby Vrysses. Its shepherds drove new black Nissan pick-ups into the hills and slaughtered fifty lambs every Easter. Only the old men of Pano called it 'Anissari the Bad' but still they visited every day to meet friends for coffee.

A single paved road skirted the basin, curving between the two villages, and at its midpoint was 'my' Anissari. Messa Anissari. Meaning 'In-between'.

Messa Anissari was neither bad nor wealthy, snobbish nor particularly hard-working. None of its buildings was of architectural merit. Its sons had not become shipping magnates, its daughters had not married television presenters. Its dozen houses straddled the loop of road, keeping no secrets, attracting no foreigners. The village hadn't rooms for rent, a rural crafts museum or a snack bar. There was nowhere to buy olive oil or wine because no one needed to buy it. Every family produced their own supply. Foreigners who came to Crete in search of enchanted bougainvillaea lanes and hidden flagstone courtyards only paused for a quick Scandal ice cream on Yióryio's leafy terrace before hurrying on. They coveted period features, not reinforced concrete. They didn't want to listen to Polystelios and his wife Aphrodite quarrel all night, or to have their own arguments overheard. Once, a dinner plate, thrown in anger by Aphrodite, had sailed clear across the road and through an upstairs window to strike Iánnis as he masturbated over a puckish issue of *Shagman*. His mother's canary never again sang when spicy *loukániko* sausage was served for supper.

'I say to my only God I love this place,' Yióryio told me while sweeping out his *kafeneion*. He wore a bleached white singlet and leaned against the broom. 'I know here I am in paradise.'

He and his roomy, rose-pink wife Sophia, always busy and good-humoured, spent their days surveying the street from behind the bar at the centre of 'In-between'. They hailed passing *mechanés*, stopped farmers to chat, shared news of tomato prices, dowries and adultery: Eleni was still carrying on with that man at the plastics factory; Zacharis, away with the army in Lesbos, spent his days killing flies; milky *flómos* sap was the best remedy for treating warts.

'*Ela*, Dimitri,' Sophia called out to the fishmonger, firing off her shrill chatter, 'Vassili will grow mangoes in the spring and no one is to know.'

From time to time the village came to a complete stop, with two or three cars blocking the road, their doors ajar, coffee brewing and Sophia leaning into drivers' windows, one after another, her ample rump stuck out into the street like a beachball bollard. Not that there was ever much traffic to disrupt – apart from Theo, the village boy racer with Tin Tin hair and a blond quiff, who sped his motorcycle to Vrysses four times a day to drink frappé with the Hotel Utopia's pretty receptionist.

'*Ela, keratá*,' Yióryio cried to Polystelios. *Keratá* meaning cuckold, an offensive endearment with an edge of truth. 'Why is gossip like Viagra? Because they take a little thing and make it this long.'

To its inhabitants Anissari was a self-sufficient, unchanging, ideal community. As they kept telling me.

I returned to Gavalohori to collect my wife Katrin. We drove over the hill and moved into the old house behind the *kafeneion*: fire in the grate, stone floor, white walls four feet thick. And as cold as a bone. The sparse building formed part of the original village, tumbledown and detached from the new, cluttered concrete boxes on the road below. Its windows opened onto the Lefka Ori and the scent of lemon trees.

'It's beautiful,' said Katrin, biting her lip. Meaning a beautiful place to live. Not to die.

Between the showers we unloaded my duffel bag of tools, wire cutters and pinking shears. I unpacked my cycle helmet and a pair of protective decorator's goggles to be worn on the flight. There were books too: Bulfinch's *Myth and Legend*, Murchie's *Song of the Sky*, Langewiesche's *Stick and Rudder* and *Aerocrafter: The Complete Guide to Building and Flying Your Own Aircraft*. I put by the bedside a schoolboy Greek *History of Flight* and Trevor Thom's *Air Pilot's Manual* ('The basic training aeroplane consists of a fuselage to which the wings, the tail, the wheels and an engine are attached').

Sophia appeared at the door, her thickset figure silhouetted against the clear, blue light, bearing a platter of chicken with lemon rice.

'And so you have come to us,' she crowed, overflowing with pride and hospitality, asking our age, my income and why we had no children.

'How much will your aircraft cost?'

Most days she would bring us simple, fresh dishes, the produce of her kitchen and garden: tomatoes warm from the sun, spilling their seed on bread, tiny *atherina* deep-fried and eaten whole an hour after it had been swimming in the sea, chips sprinkled with thyme and olive oil as green as cat's eyes.

'Ask me for whatever you need,' she said then gazed at Katrin's trainers. For all her generosity Sophia coveted other women's shoes, perhaps because hers were bought from the travelling cobbler, perhaps because her husband wore patched boots tied together with string.

I wanted to learn about the village's history, the thin thread of continuity which wound through the ages, and began by asking her when the house was built.

'Before my grandfather's time,' she said, pleased to oblige, gesturing back in time with her stout, chipolata fingers.

'And the ruins on the hill?' which I guessed were three thousand years older.

'Before my grandfather's time.'

Yióryio, cocksure and shaggy, strode up from the bar with another flask of *tsikoudiá*, this one a gift from the mayor.

'I will tell you all stories of this paradise place,' he promised, 'in names of our fields and rocks in between and where my father planting first wild figs.' He raised his glass. '*Etsigya!*' he toasted us, resting his hand on my shoulder. '*Yiá na skotósome to mikróvio.*' One more drink to kill the microbe.

I was no Hellenist. I knew no ancient Greek. My eclectic grasp of the modern language had been picked up like pretty stones on Aegean beaches during holidays. Katrin was the better linguist and would act as our interpreter, even though there was much in our venture of which she could not let herself speak. She understood my obsession. But I chose to ignore her anxiety. Grief blinded me to her and the villagers' generosity. I saw only my need to reach back to beginnings. To discover the extent to which the ancient still touched the modern. To find constancy beyond death.

Outside the house the air was steeped in wild peppermint. Shy winter crocuses appeared under the walnut tree. The garden was woven in bird song, plaited with crowing cockerels and the rush of sparrows hunting for nesting sites under the eaves. CDs suspended above Yióryio's vegetable patch failed to scare away scavengers. Tinny music crackled from Kóstas' chicken shed. Papoos, crippled and bent, trimmed his vines. Ulysses strode away from the village, the silent transistor pressed to his ear. The bread van tooted its way past him, disturbing a flock of pigeons which wheeled above their roost, their bellies flashing white against the grey stone hillside.

I hailed the van and bought a small black loaf. *Micro mávro.*

'You're the one building an aeroplane,' said the baker.

'Did Yióryio tell you?' I asked.

'I heard it in Vamos.'

Five kilometres away. Gossip spread between the villages through families, neighbours, across olive groves, by way of mobile traders who sold loaves, hoes or roasted chickpeas.

'News reaches every corner of Crete in two days,' he said. 'One day if it's interesting.'

After we'd unpacked I took Katrin next door to meet the neighbours. Polystelios was scything grass along the meagre path between our houses.

'Alexander the Great named every village in Crete according to the alphabet,' he said. Word had reached him of my interest in history. 'First Anissari, then Vamos – which is spelt with a B in Greek, Candia, Drapanos, Exopoli and on and on.'

'Alexander didn't come to Crete,' I said.

'I heard you knew history,' he scoffed. 'He came here to fight the Communists.'

More than two thousand years separated Alexander the Great and Karl Marx.

Polystelios jerked the fodder onto his back and led us into his yard shouting, 'Tell the *pilótos* that Alexander hated the Communists.'

Aphrodite emerged from the shadows of their tiny, stark hovel pushing her aluminium walking frame, moving her swollen grey limbs with difficulty. Her hips were so broad that she appeared to be wearing a lifebelt. A pox of black moles crossed her wide face. She looked hideous, cinder-eyed with a wan, bloodless complexion, yet on seeing us she pushed back a strand of chestnut hair with a knotted hand and beamed in welcome. 'What about Alexander?' she asked her husband.

'That he named the village.'

'Before my grandfather's time there were two brothers named Anissari who divided the valley between them,' Aphrodite told us. 'The good brother was given the best land for Pano Anissari. The bad brother took the land with the almond trees.'

'That's not the right story,' barked Polystelios.

'It's the one I know,' she said and started stroking my arm. 'You are too thin. Doesn't your wife feed you?' she asked with a pockmarked grin at Katrin.

Aphrodite heaved herself into her filthy kitchen, dragging me by the hand. A lean cat jumped down from the paraffin hotplate. Gaunt chickens ran free, roosting on the open stairs, their scratchings and feathers falling between the floorboards and onto the earthen floor trodden as hard as marble. On the table lay what looked like wrinkled elephants' tails.

'We've already eaten,' said Katrin, her eyes on the oily appendages. 'Sophia brought us chicken for lunch.'

'*We* are your neighbours,' insisted Aphrodite, handing us each a fork. '*We* must feed you.'

'And you must drink my wine,' said Polystelios, reappearing with a greasy plastic cola bottle. He poured out the pungent, acrid liquid. 'Yióryio's wine is like water,' he declared, 'but mine has body and no chemicals.'

If his wine had body then the body had rotted, along with the worms. It tasted like rancid dust and smelt of insecticide.

'You must visit us always,' insisted Aphrodite, dropping herself into a chair. Polystelios pushed me into the seat beside her as she diced the tails. 'You will eat with us. You will learn to speak better Greek. And I will teach your wife *pythia*,' she proposed, not taking her eyes off me. 'You understand *pythia*?'

Pythia was the name of the priestess at Delphi who delivered the oracles.

'*Pythia* is life wisdom,' she said. 'I can show her how to cook on a wood fire, how to make charcoal. Young people, like Sophia, they don't know it. They only turn on the microwave and buy new shoes.'

'Where there is laziness the devil has his workshop,' said Polystelios. The walls were bare but for a fat-splattered icon.

'I will also teach you about love,' she croaked, emitting a sound not unlike a death rattle.

'Love?'

Aphrodite and Polystelios had no children.

'It can be a quiet life in this village which is why we both have sweethearts.'

'I have two,' smiled Polystelios as he soaked a hunk of bread in the jellied juices. 'One soft and the other as hot as a gipsy. They are sisters living in Vrysses.'

'My lover is in Exopoli,' Aphrodite declared with a wink to Katrin.

Polystelios slapped his thigh in delight at the idea, or perhaps the wine was inducing involuntary convulsions.

'I run to him every time Stelios goes out into the fields,' she cackled. 'This walking frame is just for show.' She flared her nostrils as if to catch her breath. 'Now eat.'

The elephant tails were sheep guts. Polystelios and Aphrodite couldn't afford to eat their animals – apart from the intestines. Every time a lamb was slaughtered, they sold its meat, dug its entrails into the vegetable garden and kept for themselves only the tail-like *coquoretsi*.

'You're not eating,' said Aphrodite to me.

'Don't you like my wine?' asked Polystelios.

'Your wine is . . .' I searched for the right word, '. . . astonishing.'

They were dirt poor, forced to sell most of their produce, yet thrusting on us all that remained.

'Aphrodite is from Pano Anissari,' he confided in me, 'so she appreciates quality.' He refilled my glass and reached to shake my hand again. 'You are welcome in this house, *pilóte*.'

'How long have you lived here?' I asked him.

'I was born here.'

The kitchen had once been the stable with the bedroom above. Winter warmth from the donkey and goat had risen through the large gaps between the floorboards. Aphrodite saw me looking at the stairs.

'Stelios carries me up to bed in his arms every night, after I have been with my lover.'

'The tomatoes are sweet in Crete,' Polystelios told me, 'but are the women sweeter?'

'How will I ever know?' I asked, joining in their bawdy banter, telling them that Katrin had come to Crete to keep an eye on me.

'By living here with us,' answered Aphrodite. 'Not in Yióryio's old shed.'

'This is a good village,' Polystelios assured me, proud and generous. 'In Anissari one can get a bit of air. You will see for yourself because I will show you all of our valley.'

Aphrodite took my hand again. 'And in exchange you, *pilóte*, will take me to fly.'

In the evening the winter rains returned, driving the villagers back indoors to stoves, soup and television. As the twilight descended on the valley we pried ourselves away from Aphrodite and Polystelios. A car licked along the asphalt, bringing Ulysses home, and somewhere a dog barked with much enthusiasm but little conviction. Water gushed off the limestone uplands, making muddy rivers of winding paths. Yióryio's donkey, Rosetta, took shelter among the olives.

The isolated reaches of the Apokoronas peninsula were a place of elemental beauty, of rock and water, of wildfires and red-earthed olive groves, of fierce blue skies and churning seas. The cliffs seemed to plunge into the swells, or to be thrust up out of them; vital, potent, as defiant as the island's past. But tonight stout clouds fell with the darkness, swallowing the horizon, the valley, even the lights of the other houses. A stone grey sky fastened itself around the edges of the village. The bells of Agios Ioannis rang a muffled hour, too indistinct for me to count.

In Anissari's vineyards and fields the seasons set the rhythm. As they always had. Here settled society and history, in its various versions, stretched back in time further than anywhere else in Europe. Everything that could happen had already happened. Or so I would learn.

4. Flight, and My Part in Its Discovery

All my life I have been moved by aeroplanes.

At four months old a silver-bellied, three-finned Super Constellation flew me from Vancouver, where I was born, to Toronto, where we would live. A year later I crossed the Atlantic on my mother's lap in a Bristol Britannia Whispering Giant, the first commercial aircraft to fly non-stop between America and Europe. When I was eight my father took me to Washington on a TCA Viscount to gaze at the Wright brothers' Kitty Hawk Flyer. The Flyer's historic hop – man's first powered, heavier-than-air flight – had lasted twelve seconds. On our return journey the Viscount captain invited us into the cockpit and gave me a tinny Junior Pilot badge which, almost forty years later, I would take to Crete.

I launched balsa wood gliders from the attic. My Airfix Meteor skimmed the rooftop. I baled out of my bedroom window onto piles of pillows wearing an RCAF pilot's jacket.

I designed a self-propelled, two-stage paper aeroplane by rolling between my fingers small firecrackers until their plugs fell out, then mounting them in pairs in line beneath each wing. The first stage ignited the second and the glider shot forward across the garden in fitful flight. A later prototype was eaten by the dog.

At twelve years old I shivered above the Arctic in a breezy DC-3. At thirteen I laughed out loud when a float plane lifted me up in a rainbow of spray from the Muskoka Lakes. In 1967 I drank my first glass of wine on my first Big Jet, a BOAC Boeing 707, then fell in love with my first stewardess. She wore white gloves, a waisted navy uniform and winged pillbox hat. For the last winters of his life my father decamped to the Bahamas to escape the Canadian cold and the family flew south on stretched DC-8Ls to join him for Christmas.

When I moved to Europe I began to catch aircraft like buses. Twice a year I was carried home by Freddie Laker, PEOPLExpress and on the maiden Virgin Atlantic New York flight. Two dozen times I rattled down the Berlin Air Corridor at 10,000 feet, far below the PanAm 737's ideal cruising altitude but within range of Soviet anti-aircraft guns.

I lost an engine above Rangoon, crossed the Pacific on bankrupt Continental and once had a Qantas jumbo – designed for 456 passengers – all to myself (but only from Melbourne to Sydney). I lost a lover in LAX and found another at Gatwick. My first travel story won a competition and a flight on Concorde, delivering me to JFK an hour before I'd left Heathrow. Above Seattle I dined on 'a collation of smoked salmon, sevruga caviar and prawn sushi' and brown-bagged it on an Ilyushin 86 in an ice storm near Minsk. And every aircraft I boarded, from a Fairchilds Pilgrim to the Airbus, across tropical tarmac or cattle-like through a docking bay, thrilled me: the click of the seat belt, the start of engines, the surge of power, the hurtle down the runway, the anticipation of the miracle.

'More than any other thing that pertains to the body,' wrote Plato, the wing 'partakes of the nature of the divine. Its natural function is to soar upwards and to carry that which is heavy up to the place where dwells the race of gods.'

In part flying was an escape from the mundane. Why shiver in Scarborough when you can jet off to Samarkand? But for the greater part my flights were a running towards points of love. With wings I could reach out to old friends and unmet lovers, discover new worlds and fly home to my mother. Air travel was for me an emotional journey. I flew for love and flying was divine.

Daedalus had built wings so that he and his son could be free, overreaching himself to prove his vitality as well as mortality. The Greek scientist and philosopher Archytas 'invented a pigeon which could fly, partly by means of mechanism and partly by the aid of an aura or spirit'. In Christianity the faithful reached beyond their grasp to life everlasting. In the psalms man prayed for release from earthly iniquity and the terrors of death: 'Oh that I had wings like a dove, for then I would fly away, and be at rest.'

The wonder of flight transcended the millennia. In the thirteenth century the English philosopher Roger Bacon envisaged a 'flying machine in the middle of which a man could be seated and make an engine turn to activate artificial wings that would beat the air like those of a bird'. To rise on high for the glory of God.

Two centuries later Leonardo da Vinci, who knew Bacon's *Epistola de secretis opeibus*, filled his notebooks with minute observations of birds' spiralling flight. He designed a butterfly-winged ornithopter, a spearhead hang glider and a helical screw helicopter. Alongside a drawing of a caged bird he scribbled 'the thoughts turn to hope'.

Leonardo worked in secret, maybe building his *ornitottero* (with a twelve-metre wing of sized silk and pinewood strengthened with lime), maybe even risking himself in the air. But his attempts to achieve human flight failed. 'Because of their ambition,' he wrote, 'some men will wish to rise to the sky, but the excessive weight of their limbs will hold them down.'

Other would-be aviators relied more on hope than on invention, throwing themselves and caution to the wind. Giovan Battista Danti of Perugia in northern Italy crashed his flying machine into a church roof in 1503. João Torto, a Portuguese barber and braggart, fashioned fabric wings and cushioned landing shoes and launched himself off Viseu's cathedral tower in 1504. He landed on the lower roof of the chapel, lost his footing and fell to his death. An English steeplejack named Cadman the Aviator came to a similar end. In 1739 he flew from the top of St Mary's Church, Shrewsbury wearing a bird's head helmet.

> 'Twas not for want of skill
> Or courage to perform the task he fell;
> No, no, a faulty Cord being drawn too tight
> Hurried his Soul on high to take her flight
> Which bid the Body here beneath good Night.

Cadman may not have lived. His body may have been crushed on the damp stones. But his spirit found wings.

With the Enlightenment the dream of flight became a technological challenge. Analytical thought swept aside medieval hubris. Inspired by the smoke they observed curling up a chimney, the practical, paper-making Montgolfier brothers constructed the first hot air balloon. In 1783 their second flight, exhibited for the king at Versailles, carried the first air travellers aloft.

'The stately structure, which was gorgeously decorated, towered some seventy feet into the air and was furnished with a wicker car in which the passengers were duly installed. These were three in number, a sheep (named Mont-au-ciel), a cock and a duck, and amid the acclamations of the multitude, rose a few hundred feet and descended half a mile away. The cock was found to have sustained an unexplained mishap – its leg was broken – but the sheep was feeding complacently and the duck was quacking with apparent satisfaction.'

In 1798 the Jesuit father Lana conceived a dainty air boat, held aloft by four balloons, in which the aeronaut sat at ease handling a little rudder and simple sail.

Six years later a Yorkshire baronet, Sir George Cayley, laid the basis of the first successful aeroplane – a free-flying, fixed-wing glider. In his pioneering work he defined the principles of heavier-than-air flight and the forces acting on the wings. Cayley convinced his coachman to be the test pilot but, after a single flight, his man refused ever again to be parted from the earth: 'Sir, you engaged me to drive your horses,' he pointed out, 'not to fly.'

Then in 1857 a Frenchman, Jean-Marie le Bris, mounted his glider on top of a carriage and instructed the driver to set the horses to gallop. Le Bris took to the air but a rope caught around the seat of the carriage, carrying it and the driver aloft. They both came to earth unharmed, achieving the first recorded two-man flight.

Towards the end of the nineteenth century the German Otto Lilienthal made over two thousand flights in bamboo and waxed cotton 'soaring' machines, the first hang gliders, refining the system on which the Wright brothers would base their Flyer.

'It is a difficult task to convey to one who has never enjoyed aerial flight a clear perception of the exhilarating pleasure of this elastic motion,' he wrote. 'The elevation above the ground loses its terrors, because we have learnt by experience what sure dependence may be placed upon the buoyancy of the air.'

His sure dependence did not save his life in 1896 when he crashed and broke his back.

Next to defy the earth-bound was Alberto Santos-Dumont, a dapper Brazilian dandy who flew airships around the Eiffel Tower at the start of the twentieth century. He lifted off from his house near the Champs-Elysées and once crash-landed in the Baron de Rothschild's Parisian garden, where he was served lunch in a chestnut tree while being disentangled from the branches.

A slight man of fierce determination, Santos-Dumont recalled another crash, 'For the moment I was sure that I was in the presence of death. I will tell it frankly, my sentiment was almost entirely of waiting and expectation.

"What was coming next?" I thought. "What am I going to see and know in a few minutes? Whom shall I see after I am dead?"

'The thought that I should be meeting my father in a few minutes thrilled me. Indeed, I think that in such moments there is no room either for regret or terror. The mind is too full of looking forward. One is frightened only so long as one still has a chance.'

As a child *The Boy Mechanic* ('1,000 Things for a Boy to Do') had been my favourite book, as it had been my father's before me. On its green cloth cover a clean-cut young American, wearing a tie and workshop apron, rolled up his sleeves to finish a model monoplane. Published in 1913, its pages were packed with practical plans for improbable devices: Build a Snowball Thrower, Make Foot Boats to Walk on Shallow Water, Electrify Your Garbage Can to Shock Away Scavenging Racoons. The book recalled a resourceful, self-reliant age when idle moments were spent producing rather than consuming. Over the years I had browsed through its plans for home-made Telegraph Sounders and motorcycle-powered washing machines. With every reading one article always caught my eye: How to Make a Glider.

'A simple glider of the monoplane type can be easily constructed in a small workshop,' stated the text. 'The costs of construction are not great and the building does not require a skilled workman.' The book's author claimed that any boy could build a flying machine, 'as long as his wood was straight grained and free from knots'.

Not only that, but he could fly it himself.

'Take the glider to the top of a hill and step into the main frame. Face the wind and run a few steps. You will be lifted off the ground and carried down the slope.'

Easy.

On our Anissari balcony I toyed with the idea of making a glider until Katrin said, 'If you jump off a cliff without an engine I'll be a widow.'

An engine added complication but in truth I was happy neither with the idea of throwing myself off a Cretan cliff nor with being dragged into the air by a tow-car. My aeroplane needed to be self-propelled.

So I began to look beyond *The Boy Mechanic*, though not beyond its 'stick and string' decade of manned flight. The aircraft of the years before the First World War appealed to me. Flight was then a spiritual adventure as much as it was a technological event. For early aviators like Antoine de Saint-Exupéry the aeroplane was 'an instrument for human growth'. Flying machines heralded a new age, with early aviators realizing 'the elusive dream that man has cherished from far distant days'.

Today aeroplanes can be bought in kits and assembled like Ikea kitchens. My *AeroCrafter* guide buzzed with home-build Flying Fleas, Early Bird Jennys and Viperjets. The RagWing Sport was sold with a whole-aircraft parachute. The Legal Eagle came complete with fixed pitch propeller and shopping cart tail wheel. But buying a kit was for me both expensive and a bit like cheating. I needed to follow my own instruction book and to build my machine from nothing, or at least from bits and pieces found in a Cretan village. Not by mail order from Oshkosh, Wisconsin. I didn't want to be reliant on prefabricated, composite materials. I had to depend on myself alone, rebuilding myself as I built the aeroplane. But I wasn't a fool. I accepted that I knew nothing about aerodynamics. I needed a plan.

I began to look for a simple, classical design which I could cobble together and which wouldn't fall out of the sky.

Over glasses of Yióryio's retsina I considered the Antoinette and the Blériot 22, Farman's biplane and Roe's triplane. All of them looked fragile and complicated. The White monoplane was a paper-thin wisp of a machine with less substance than a mathematical diagram. Dormoy's Flying Bathtub looked like an airborne coffin. The 1911 Gaunt Cycloplane Flying Bicycle did not lift my spirits.

Then I found the Demoiselle.

In 1906 Santos-Dumont turned his attention – and considerable fortune – towards powered, heavier-than-air flight. In that year his 14-bis, a puffed-up backwards box-kite, hopped 722 feet across the Bois de Boulogne. He wore a dashing Panama hat for the flight. The undercarriage collapsed on landing. The charmed aeronaut was carried away in triumph.

But his second aeroplane, the dragonfly-light Demoiselle, was the greater invention; a fifteen-foot-long bamboo pole with an aero engine at the nose, wings in the middle and an empennage tail. The pilot was slung between the wheels on a leather seat no more than six inches above the ground. The Demoiselle was a beautiful flying machine – small, fast and inexpensive – and the forerunner of the modern light plane. In its time it was so popular, despite its peculiar warping wings, that Santos-Dumont licensed production to the Clement Bayard Motor Company and 'placed his invention at the disposal of the world in the interest of the art to which he has devoted his life'.

I imagined that it would be easy to find a set of the plans and from Yióryio's *kafeneion* telephone called the United States. Scott Perkins was 'Director, Historian and Preservationist' of the Vintage Ultralight Association in Marietta, Georgia. He knew the Demoiselle.

'It's a tough project,' Scott told me in a Southern drawl, after confirming that he had the technical drawings. 'You have to work with bamboo.'

'Bamboo grows here on Crete,' I shouted at him. Behind me Kóstas the policeman and Papá Nikos were arguing about a fall in the Athens stock market.

'What exactly are you trying to do?' he asked.

I explained.

'Do you know a magazine called *Popular Mechanics*?'

I did. *Popular Mechanics* had been the mainstay of the amateur American handyman for over a century. *The Boy Mechanic* was one of its early publications.

'Every month *Popular Mechanics* is chock-full of goofy offers designed to take your money, things like hovercraft wheelbarrows and engines that run on water. Back in 1981 they ran an article on a plane that anyone could build over a few weekends.'

'Did it fly?'

'*That* was the goofiest thing of all. It flew just fine – because its design was based on the Demoiselle – but not too far. It was called the Woodhopper.'

I liked its name: direct, honest, manageable.

'There's a guy living on a lake near here who built one with floats. Every morning he hops to some quiet bay, puts down and starts fishing as the mist is rising off the water. Man, you cannot have more fun in life, even if you own a P-51 Mustang.'

'Do you have the Woodhopper's blueprints?' I asked.

'You bet.'

'Can I buy a set?'

'As long as you're not suicidal or have a terminal illness.'

I asked Scott to explain.

'If you buy these plans you should definitely not follow them. The original design was way too light. You will need a stronger engine, heavier wood, thicker wire, better everything. And if you do not heed these words you will die.'

I got the message, ordered the plans and asked him to courier them to Greece.

'Sure thing. Have a good day now.'

I ran back to the house to tell the news to Katrin. Polystelios was skinning a goat and Aphrodite sat in the lane watching her neighbour's television through the open door. She waylaid me with a glass of *malotyra* tea.

'Did you tell Scott that you're not a pilot?' Katrin asked when I reached home. It had slipped my mind. 'I keep thinking that you don't value your life any more,' she sighed.

★

When the plans arrived two days later I unrolled them in the *platía*, the whitewashed village square no larger than a room and domed with mulberry trees. Katrin weighed down the corners with ashtrays and an ouzo bottle. The men pushed around for a better look. Iánnis said, 'I saw this on television,' and sat back down in the *kafeneion*.

Scott had sent me six large blueprints: an overall view of the Woodhopper's structure, a cross-section of the wings and tail, a full-scale wing rib template, a spider's web schematic of the structural wires and two sheets detailing incomprehensible aluminium fittings called king posts and control horns. Across the top of each sheet was printed a warning: 'These drawings are for engineering, reference and study purposes only. Structural analysis and safety data are not available and may never have existed.'

'Where will you find the engine?' asked Papoos, wasting no time getting to the point. He – like the other men – appreciated machines. Machines were modern, made noise and cost money.

'I don't know yet,' I said.

The accompanying Construction Manual made light reading: a slight, photocopied, fifteen-page document which included a section on 'Learning to Fly' ('Always wear a helmet. Do not taxi or take-off without fastening your seat-belt'). Scott had sent a copy of the original Build-the-$900-Airplane-Anyone-Can-Fly article. It too was short on detail. My first impression was that much would have to be improvised as I went along.

Yióryio refilled the glasses, distributed plates of olives and walnuts to 'cover' the alcohol, then slumped into a chair. He and Sophia were like the flowers, either fresh as daisies or wilted like dried roses. This morning, having kept the bar open late the previous night, his head hung low.

'Here's the shopping list,' said Katrin.

The list of materials ran to less than half a page: one length of good Douglas fir, seven sheets of polystyrene, cable, glue, aluminium tubing, hinges and dress-lining fabric, bicycle wheels and a bleach bottle petrol tank. There wasn't much to the Woodhopper.

'Where sitting the passenger?' asked Yióryio, pointing at the plans.

'It only carries a pilot.'

'*Panayia mou*,' said the priest. All Holy Virgin. 'This is short-sighted. You must make two seats to take people and profit from your labours.'

I spread a set of photographs across the ground. There were shots of the Woodhopper at rest, taxiing, lifting off and rolling into a steep turn. In each of them the aeroplane bore a strong resemblance to the Demoiselle. The broad wings, slender body and nose-mounted engine gave it the look of a giant, wispy dragonfly. The pilot sat among its wiry legs, his feet on the axle, unprotected by cockpit or windshield. It wasn't a complicated design. It also wasn't a few weekends' work.

We deciphered the plans as the baker stopped by for his morning coffee and played a hand of '66'. Yióryio's aunt led a goat on a rope along the street. In the cool sunshine I felt a tentative swell of optimism. Katrin was less impressed.

'This Supplementary Sheet corrects the errors on the blueprint,' she said. 'There are dozens of changes.'

According to the specifications the Woodhopper weighed less than I did. Its wingspan was 32 feet. It took off at a mere 27 mph, cruised at 35 mph and had a range of about 30 miles (depending on the size of the bleach bottle). I searched the plans for details of the engine. The few references that I found were vague. 'If power is too low take-offs will be touchy and more susceptible to ground loops.' There were no clues on how to determine if power was too low. Or what a ground loop was.

'You can borrow my rotavator's motor,' said Socrates, which was generous of him but of limited aeronautical use.

'And my compressor,' offered Yióryio, not to be outdone.

'Do you remember when your father brought the first rotavator to Anissari?' the shepherd asked Yióryio.

'*Your* father made him eat wood.' Meaning he beat him.

'He feared for his job.'

'A field then took twenty men to dig in a day,' remembered Papoos, opening his hands as if to show off his calluses.

'And now Anissari will have the first aeroplane to be built on Crete,' I said.

'Not first,' said Yióryio. 'My uncle built first.'

'Your uncle?'

'I will tell you the story,' he said, rousing himself to the task. 'In war many aircraft come in the Crete.'

There had been Spitfires, Hurricanes and, after 1941, Stukas and Dorniers.

'My uncle seeing them and decide to carve his own. He pull maple log from river with donkey.'

'A carved aeroplane?' I asked. 'Did it fly?'

'*Kala re, Amerikanaki eisai?*' What are you? A young American? That is, naive. 'My father hit him – because he should have been planting watermelons – and called him *cocaliáris*.' Meaning bony, an airhead.

'*Cocaliáris*,' repeated Iánnis, laughing to himself.

'Then this will be the first aeroplane to be built and to *fly* on Crete,' I conceded.

Kóstas the policeman arrived in his bulldozer to distribute the village water bills. The baker continued with his deliveries. Papoos slipped away to trim his vines.

'I can't find the bolt page,' said Katrin, shuffling through the plans. 'It's mentioned here on page seven but I can't find it.'

'I don't suppose there are any engineers living in the village?' I asked. I'd noticed another warning on the blueprints: 'If an actual project is planned qualified engineers MUST be consulted.'

'I know one,' said Yióryio and the other men laughed.

'I could do with his help,' I said.

'*Her* help,' said Iánnis.

'And you mentioned last night that you could find me a workshop,' I reminded Yióryio.

'I tell you it is no problem,' he said, scratching his stubble and frowning. It would be an insult not to accommodate me. 'But is time now? Or later?'

'Whichever you prefer,' I answered, unsure how much to push him but anxious to begin work.

'Now or later I find time,' he assured me, relieved not to be committed.

'Later is better?'

'Of course. Just now I must go to give food to my animals.'

'Tomorrow morning then?'

'Morning or afternoon we go tomorrow,' promised Yióryio. 'Or after tomorrow.'

I began to gather up the plans. Santos-Dumont didn't teach himself to fly in a day. The Wright brothers took nine years to build the first practical aeroplane. The rushed, maiden flight of the 1912 Passat Ornithopter, a cunning, cigar-shaped, flapping machine, ended when 'its passage was arrested by a tree' on Wimbledon Common. I, like my predecessors, had to learn to be patient.

5. Winging It

We went 'after' tomorrow. Four days after. Yióryio had more pressing responsibilities; his nephew needed a lift to Sitia and on the way back the exhaust fell off his weather-beaten pick-up. He spent the night with a relative near the garage in Hania then – as he was so close – drove to Horasfakion to help his brother-in-law prune olive trees. The brother-in-law then borrowed the truck to visit his sister who was having a baby in Vouvas. The exhaust fell off again near Petres. When Yióryio finally returned home Rosetta was off her food and had to be taken to the vet.

In the end we arranged to meet in the square at eight o'clock the next morning. I arrived as the sky was taking on light. He was nowhere to be found. I drank a coffee, ate a *bougátsa* pastry and walked back home. At eleven I returned to the *kafeneion* to find that he had come and gone. I drank another coffee.

'We are human,' shrugged Sophia and asked me if I was free tomorrow. I walked home again.

At noon Yióryio shouted up to the house. 'I will show you my land,' he yelled. 'It is five times beautiful and making you jealous.'

Yióryio believed that he could charm birds out of a tree. A gun would be faster.

In the sun-faded pick-up we set off along the rocky shoreline. His exhaust sounded worse than ever. To the left silver waters reached to the horizon. On our right a line of razor-edged peaks cut across the sky. At a craggy crossroads Yióryio turned inland and climbed fast up a mountain track which wound like a small intestine into the body of the island.

In the ten days since our arrival spring had seized Crete. The winter rains had eased and lime green shoots had appeared on the vines. The dark hills were dressed in wild almond blossoms which smelt of marzipan.

'I love this smell,' beamed Yióryio, inhaling deep breaths of air, plucking an insect from his moustache. 'I cannot be away from it.'

I smelt thyme, mint and the salt of the sea. Cypress and oak towered above thickets of arbutus and tree-heather.

'That tree I remember from when I boy,' he said, pointing at a thick-trunked olive. 'And those stones show beginning my cousin's dowry.'

I hadn't been prepared for the Cretans' *páthos* – brute passion – for their island.

'Here in paradise I have chicken and sheep. I hold one hundred trees to make oil. I grow leek and potato. I give no pills to my animals with bones on package.'

Skull and crossbones, the poison symbol.

'Other farmer put chemical on vegetable then get cancer.

Or buy machines and spend time chasing after paper. I am not this farmer. I dig with old fork to sweat out the *raki*. In the winter when I am cold I cut wood. My mother make cheese and yoghurt. Rosetta my donkey is like a flower. This is all I need.'

We skirted above the ruins where we had met and Apokoronas revealed itself as a province of small fields, of stepped terraces that dated back to the Bronze Age and of dry-stone walls topped with spiny bushes, as in Homer's time. Along the roadside there were shepherd's round stone *mitáta* and oval bee-enclosures, field houses and threshing circles. A ribbon of plane trees lined a tangled ravine. Across the valley Kóstas' bulldozer clawed a new road out of the maquis.

I asked Yióryio if he ever worried.

'My fear is I am too good,' he replied, laughing out loud. 'Because Big Boss is in me, thanks to God.'

'I mean death,' I said.

'This morning in *kafeneion* men worry for no rain. Maybe it rain tomorrow, and then after two days raining men will worry for too much. So why please? We know nothing for tomorrow.' Then he asked me, 'You believe in Big Boss upstairs?'

'I don't know any more,' I admitted.

'I have many receipts,' he said.

'Receipts?'

'Like in supermarket. You buy a kilo of honey and woman give receipt,' he explained. Proof of purchase. Or debt paid. 'When my boy Leftéri was eight months he catch meningitis. I go hospital with my son in my arms, his head hanging. Doctor very sad and put wires on him. Then that night I see dream of white flower which lose all petals but one. In morning I go to the hospital and my son is better. They take off all wires but one which will not come, like petal. I never forget my dream.'

Yióryio took his eyes off the road and stared at me, nodding with sage-like wisdom. He reached out through the story and in return I couldn't even summon my compassion. We rattled down a dirt track, crossed a carpet of lemon yellow oxalis daisies and stopped.

'This my land,' he said, moving on, stepping out of the pick-up, embracing the patch of wild ground. 'You can use for three months for no money. I want no money.'

A faint whipping of wind snapped through the prickly oak. There were black nets folded in the crooks of the olives and artichokes planted under the orange trees. It was an attractive spot and his offer was kind but it was nowhere to build an aeroplane.

'I need walls and a roof, Yióryio.'

'No problem, my friend. We build warm house here.'

'A warm house?' I asked.

'A polytunnel.' The long, plastic greenhouses stretched across much of the island. 'I get money from Europe and grow tomatoes after aeroplane finished.'

'Yióryio, it will take months to get a grant.'

'No more than six. Or nine.'

'That's too long to wait,' I said.

'Then I show another place,' he shrugged and stepped back into the cab. 'I am happy.'

In Apokoronas, as in most of Crete, the farms were small. Agricultural buildings were compact, suited for modest harvests and single families, not intended for use as aerodromes. I had no idea where to find a hangar.

In eucalyptus-shaded Pefki we met Stelios, Yióryio's dark-skinned cousin. Stelios had once owned a tiny jewellery shop in Hania. He had sold it, returned to his village and used the money – with matching EU funding – to build a village hall, which he now treated as home.

We sat together in the open concrete space drinking ouzo and cracking peanuts. Yióryio explained my needs, speaking in Greek too fast for me to follow, while I measured the building from card table to kitchenette. The Woodhopper's thirty-two-foot wingspan would fit, if the windows were left open.

'This is second good place I find you today,' Yióryio told me.

'Yióryio is my best cousin so the room is yours . . .' Stelios said. His only concern was that Pefki's elder villagers – who joined him in the evening – might not know what to make of an aeroplane. '. . . As long as you dismantle your machine every night.'

Next we drove down the hill to Vrysses, the market town at the crossroads of settled and agrarian Crete. In Vrysses the hills touched the plains, subsistence shepherds dressed in black Armani and drove unlicensed Cherokees. Front desk clerks read *Farming News*.

We made inquiries first at the dairy. They had an empty shed which was much too small. The manager of the olive oil factory then tried to find me a space between the vats, but the stench was nauseating. Across from the police station we spotted an empty, two-floor concrete shell. Its lack of walls was not ideal but Yióryio seemed to be running out of ideas. He asked a policeman about the building and learnt that the owner had skipped town after seducing the fire chief's daughter. One week later the half-built shell was engulfed by flames, as it would be again – we were assured – any day soon.

'Don't be so worry,' Yióryio said to me, 'or you get heart attack.'

Our last call was on Antonis Braoudakis, who ran the

unofficial Vrysses museum above a shop selling traditional Cretan products: wine fermented in his own vats, oil pressed from his olives, avocado anti-wrinkle cream made in Argiroupolis. Braoudakis broke off from his work – he was cutting out new display shelves with a chain saw – and offered to help.

'I will apply for a grant to build a polytunnel in my field,' he too proposed. 'And afterwards I can grow tomatoes.'

Our lack of success did not disappoint Yióryio. He was enjoying the afternoon, whatever came of it. He wanted to help me and I was grateful to him . . . even if he was useless.

'Every delay for a better reason,' he advised me, stepping back into the truck.

On the drive back towards Anissari I asked him, 'Don't you think someone else must have built an aeroplane before in Crete?' I had been thinking about his uncle. 'I can't believe that I'll be the first to fly here.'

'Maybe in old days.'

'In the early days of flight?'

'Of course, why not?' said Yióryio, always happy to oblige, regardless of the truth. 'In Crete everything is possible.'

'Who would know?' I asked.

'I know best,' he assured me. 'I find out for you.'

Then I saw the abandoned school.

The yellow stucco building stood at the edge of Pano Anissari. Its playground was overgrown with saxophone-mouthed arum lilies. A fig tree had taken root in its broad central steps. Discarded cars rusted under the pines.

'If it is closed we smash window,' said Yióryio. He saw my surprise and explained, 'This is only Pano.'

Thankfully the doors were unlocked.

Inside were two large, airy classrooms as high as they were long. Sunlight flooded through four tall windows. The left-hand room was empty and clean. It had a new floor. The right-hand room was cluttered with stacks of church pews and political posters. There was no woodworm or termites. There was electricity. I imagined the walls steeped in decades of children's laughter. It was a place where boys had made paper aeroplanes and girls launched spinning windmills of *askelitúra* flower heads.

A trapped goldfinch flew back and forth above the carcasses of dead sparrows. I threw open the window and waved it out into the open air.

'You like?' asked Yióryio, as if the school had always been on his itinerary.

'It's perfect,' I said.

'Good,' he assured me, 'because the mayor is my uncle.'

We turned the pick-up around and drove back down the hill. The *koinótita* was a four-room, marble and concrete office behind a memorial to Cretan patriots. As we entered, the council chamber door was hurled shut. A planning meeting seemed to be underway. We introduced ourselves to a woman at reception who was only waiting there for a friend. Nevertheless she hovered at the door while an argument subsided then announced us to the township's mayor.

He swept out of the meeting, pleased to be released, a short, pot-bellied bureaucrat dressed in a checked jacket. His soft, flaxen hair rose up and away from his forehead, following the contours of his head and fluffing over his collar.

'How can I be of service?' he asked, ushering us into his office, ordering coffee. 'Please let me help you.'

Yióryio explained my intentions. I talked about the empty school and unrolled the Woodhopper plans. The lobes of the mayor's fleshy ears drooped like his moustache but laughter stole into his hazel eyes. He seemed tickled by my undertaking.

'I have no objection to you building your aeroplane there,' he said after a few minutes. 'How long do you need?'

'I will be finished by May,' I said. With luck.

'Good, because we will use the school in June.'

There were no questions about legality. No mention of a fee. The mayor smiled and dispatched the real receptionist to find a sheet of paper. It seemed too good to be true.

'Please will you write a request for me to present to the Council,' said the mayor.

'The Council?' I asked, wary now. 'How soon do you think we might know an answer?'

'The Council will be no problem,' he replied, lowering his voice. 'We only ask them so they feel important.'

A crooked, bone-thin *yiayiá* shuffled into the office carrying a tray of small, pungent, black coffees. As I wrote out my request the mayor read over my shoulder. 'Say there that you will write in your book about our district.'

'The *pilótos* will write about everything that is true,' Yióryio assured him, 'but only mention good things.'

I made the appropriate amendments.

'Come back the day after tomorrow,' the mayor said, pleased to be asked to help. He couldn't stop smiling.

Outside on the street Yióryio swaggered, 'I am sure the mayor is laughing at you but in a good way. Like an older brother laughing at his younger brother. He likes to keep the child alive in himself.'

We climbed back into his lobster-red pick-up.

'Now I am sorry but I have to go and count my sheep.'

On Friday Yióryio brought me back to the town hall.

The Council had met and made a decision. Of sorts. At first all the councillors had approved my request to use the school, especially when the mayor mentioned the possibility of television coverage. But a lone dissenter had urged caution, reminding his fellows of the consequences of hasty action.

And of dealing with foreigners. The previous year another *koinótita* council had agreed to help a German woman to shelter abandoned animals, only to find that she was publicizing the well-known Cretan tendency to abuse dogs. Then another councillor recalled the hurried decision to erect a monument to a local family who early last century had fought the Turkish occupiers. It was discovered after the unveiling that the family had killed their Muslim neighbours and stolen their land.

In the end the councillors agreed only to put off the decision until the next month's meeting. The dissenter was asked to telephone me. I learnt much later that he was a building contractor, in charge of the work due to begin on the school in June. His call never came.

'Every obstacle is for good of project,' cheered Yióryio on the drive back to Anissari.

'I should have had a contingency plan,' I said, furious with myself for being so naive, so in need.

As we drove into the village I noticed a large white building with two heavy steel doors.

'What's that?' I asked Yióryio.

'It is two garages,' he said, driving on.

'They're big, Yióryio. Very big.'

'They are no good for you.' He thought for a minute. 'They are full of materials.'

'We could move the material.'

'The garages belong to Kato Anissari co-operative,' he explained, 'so they not let you to use.'

'Why on earth not?'

'Because you live in Messa Anissari, of course.'

I didn't understand him. The garage was on my doorstep. As my anger rose Yióryio said, 'Have I tell you about *marmánga*?'

'*Marmánga?*' I repeated, despairing of another digression. I reached for my dictionary.

'You not find *marmánga* inside any book. It comes from soul. It is something inside every one which can destroy.'

'What does it have to do with the garage?'

I had little patience now.

'When big things happen in life, you lose balance. We say for example her greed ate her. That's *marmánga*. You see?'

I saw nothing.

'With your aeroplane building, *marmánga* is waiting to let you be success or not. If you crash – pray to Big Boss it not happen – then we say *marmánga* ate you. But if you go good in your flying, then we say that you ate *marmánga*.'

'So my *marmánga* is my ambition?' I said, struggling with his dubious theory and my raw emotions.

'Or maybe your sad.'

I guessed that maybe *marmánga* was a part of an individual's nature, a modern personification of Greek tragedy's fatal flaw. Or maybe not.

'And what is your *marmánga*?' I managed to ask him.

'My *marmánga* is generosity,' he said, without humility. 'I could be rich man, with all hours I work, but every time money coming in my hand I give away – on food, on presents, until every month I don't know how to pay the telephone bill.'

Yióryio, like most Cretans, seemed to consider himself perfect. Or certainly more perfect than the rest of us.

'But I do pay it,' he went on, 'so every month I eat *marmánga*. *Marmánga* either destroy or is destroyed. Every country person understands, but lawyers and presidents don't know it.'

'So I can't have the garage because of this *marmánga*?' I said.

'You can't have garage because of *fthónos*,' Yióryio confessed at last. Jealousy.

He pulled to a stop outside the *kafeneion* before I could ask whose jealousy.

'I think you need live in the Crete for two years to understand our continent,' he told me, lips pursed beneath his whiskers. 'You drink a lot of wine, sleep in afternoon then – after eighteen months – you wake and understand.'

'I can't wait eighteen months,' I said. Bereft. Vacant.

'This is your problem.'

On the last morning in January Katrin and I drove down to the coast and dived into the cold water. The sea was a mirror as white as the sky and fringed in aquamarine, its waves rising in chuckles against the black rock shoreline, then sighing back into its depths. We swam across the glassy bay, into a day so bright that we could not see the far hills.

In this beautiful place I felt like weeping. I tried to squeeze the tears out of me, so that they might mix with the salt water, but I couldn't bring myself to cry. Instead I saw myself as a child aged six years old. Or seven. It was the evening after a forgotten neighbour's funeral. Or after a television programme. Or after my father and aunt had discussed where they'd like to be buried.

'Don't talk about it,' I had asked them.

'It's important to talk about dying,' my father had said. It is part of life.

Then I remembered lying in bed, tucked under snug blankets, my body small in the acres of sheets, clutching a toy rabbit, the favourite of that moment. Ears against my chin. The blanket's cool trim between my fingers.

In the dark bedroom I was aware – for the first time – of death. I couldn't imagine not being, a child's vibrancy locking me into the present. Rather I pictured death as simply being placed in a coffin, the body – *my* body – still alive, feeling the satin pillow, smelling the wood, seeing the last chink of light as the lid was closed and the coffin lowered into the ground. I wasn't afraid of being buried alive, or even of dying itself; I trusted that when I died my parents would do what had to be done and bury me. It had to be like falling asleep. It didn't occur to me that death could mean an end of feeling. I couldn't conceive of such a state.

I began to cry in the dark.

My mother came in and sat on the edge of the bed. Maybe I had woken her. Maybe she hadn't yet gone to bed. She probably stroked my forehead and reassured me.

'When I die,' I asked her, 'can my rabbit be buried with me?'

And she said yes; a serious response, a promise. It wasn't that I was afraid of a sleepy death, whatever it was; I just didn't want to be alone.

I drifted on the surface of the Cretan Sea and my grief, watching the waves reflect the rising sun in fierce flashes of silver. A breath of wind hatched the surface, marbling the water to the far distance. I reached out to scoop the crests in my hand but they flickered away as they neared me. I turned left and right, looking for another wave to wash towards me, then realized that the radiance was dancing in my eyes, at my feet, around the rim of my wake. I couldn't hold onto the blinding light.

We clambered back on the shore, drying ourselves and my damp, dark, northern thoughts in the sun.

6. To Make One Aeroplane

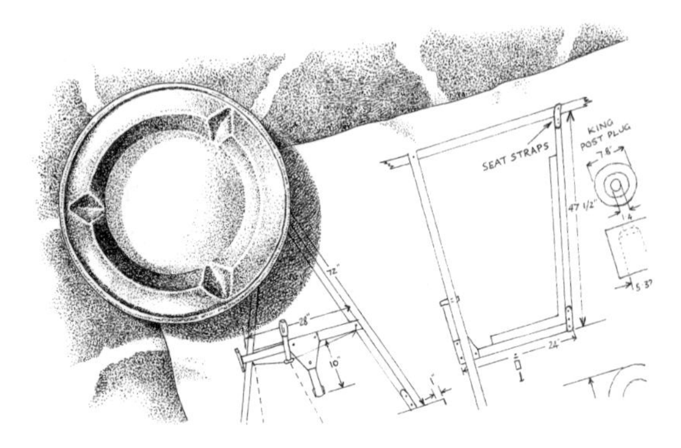

No one in Crete knew anything about building aircraft. Or so I'd thought. Until salvation came with a sharp knock on our door.

Ariadne was a compact charge of frenetic energy: sincere, candid, intense; a short radical in long skirts bearing spring flowers. She conformed to no image of Greek womanhood that I'd ever encountered. Her hair was neither bobbed nor long. She spoke with an American accent. A smudge of yesterday's make-up underlined her green – or were they violet? – eyes.

'Yióryio he said you need an engineer. I am a small engineer.' Plus mother, divorcee and orphan. Yióryio was her tall cousin, and more. 'I've had interesting days,' she said, sucking at an unfiltered cigarette. 'Later I will tell you, if I like you. Now you tell me what you will do.'

I spread out the Woodhopper's plans. She slipped on half-moon reading spectacles and flicked through the sheets. Her hands were always moving, touching, stroking, picking at fingernails gnawed to the quick.

'Flying this will be nothing like anything you've already flown,' she said.

'I've never flown,' I told her.

'Then like Icarus you will fly only once,' she said, raising her glasses in concern, 'because you will lose your life.'

'I just need to lift off the ground and touch back down again,' I said. 'Like the first Wright brothers' flight.'

Ariadne had been born into a Greece that didn't want girls as children. She wasn't pretty, hated lace-making and couldn't cook *pastítsio*. In Anissari she grew up wanting to be a boy. She filled her pockets full of stones and hung from trees to stretch herself taller. She taught herself to catch eels and to solder circuit boards. She idolized her dogmatic father, who mocked her defiance, and disdained her mother, who feigned feminine weakness. They arranged for her to marry Yióryio, the boy next door, and join their land with his father's land. It would have been a sensible, conventional match. But Ariadne denied them. She wanted the wide world, not Yióryio's shallow valley.

She left Crete and became the first woman to train as an aviation engineer at the Athens Polytechnion. At college she was outspoken, campaigned for women's rights and graduated second in her year (the prime minister's son was given the top mark). She went on to work for Boeing in Seattle designing undercarriage hydraulics and fell in love with the jet blade shop's submanager. She became pregnant. He persuaded her to return to Crete. Then not long after their arrival and marriage, the submanager eloped with a waitress who cooked the best *moussaka* in Apokoronas.

In Hania Ariadne raised their son by herself, without help from her husband or parents, with Charon waiting watchful under her mulberry tree. She earned her living maintaining Hercules transports at the air force's Souda Bay base. In the evening after her little Apostoli had gone to bed she taught herself to make – unexpectedly – animated mythical figurines.

'The dihedral is high so turns will be heavy,' she said to me, stabbing at the Woodhopper's plans. 'It won't tolerate any crosswind so you have to fly before the summer winds.'

'You're not impressed.'

'Come oooon,' she cried, jutting her jaw. Mocking. Sceptical. 'I don't usually look at disposable aeroplanes. At least the nine:one glide ratio is good.'

Whatever that was.

'Aircraft design has advanced since the Demoiselle – and this Woodhopper,' she said. 'For the amount of time you'll invest, why not buy a modern kit?'

'I like its simplicity,' I replied. 'It seems to be one of the last designs that any idiot could build by himself.'

'So tell me why you are making this?'

I told her and she folded in on herself like a collapsing paper doll.

'You have to do it,' she said with sudden emotion, pulling herself together.

Ariadne explained that both her parents – the idolized father and manipulative mother – had died at the end of last year, within fifty days of each other, twenty-one years since she had last spoken to them. She had lost her chance to say goodbye.

Neither of us spoke for a moment.

'In Crete you do not have to scratch a lot to find the truth. The varnish is not too deep.'

Ariadne lit another cigarette and, to fill the silence, asked me about a workshop. I reported on finding the school and waiting for the reticent councillor.

'But there are two big garages here in the *platía*,' she said.

'Yióryio told me that they're full. I looked last night and one of them was empty.'

'Then you can rent it.'

'He said that I couldn't work there because of *marmánga*.'

'*Marmánga?*' she asked.

I tried to explain.

'Where does he get this shit?' she asked with obvious exasperation. '*Marmánga* is a slang word, an Albanian word. It means someone who can surprise you, unlike that old devil himself. What you are describing I would call *moira*.' Fate.

I began to grasp that Greek explanations – like their myths – came in at least two versions.

'Forget Yióryio,' advised Ariadne. 'Let us find my better cousin.'

In the back of her rusting estate car was a papier-mâché bull and cackling, red-nosed Charon animated by a wobbly crankshaft. At home she said there were over four hundred other mechanical sculptures: a clockwork bonking Zeus, a compliant polystyrene Europa and a dancing Minotaur.

Katrin considered the Underworld boatman and asked Ariadne how she had become interested in automata.

'My life is lived in pieces and the puppets are in pieces,' she said.

Katrin nodded.

As we drove past the *kafeneion* Yióryio saw her car and turned away to serve a customer. Ariadne made a point of not looking at him. Then I caught her eye in the rear-view mirror.

We found Socrates, her other cousin and president of the co-operative, outside the garage leaning against his Toyota. He was watching his sheep in the hills beyond.

'You see the garage,' he said, formal in his authority, binoculars and worry beads in his hands. 'It is big and here you can make everything, including an aeroplane.'

The light, lofty room opened onto the square. It had small, high, south-facing windows and two workbenches. A five-ton truck laden with lime sacks could drive in through the sliding steel doors and an aeroplane – with wings detached – could slip out. I checked the dimensions, my steel tape measure rippling across the concrete floor and lifting puffs of whitewash. Nose to tail the Woodhopper would fill the space with only inches to spare.

In years past the garage had been used to store olives, to mix quicklime and to blend chicken feed. A decade of powder had caked its walls and ceiling in a crumbling white scale. Every breath of air dislodged a blizzard of sticky, nutrient-rich flakes.

'But why you want it for only two weeks?' asked Socrates.

I explained that I had been promised the use of the old school and I did not want a change in plans to upset the mayor.

'The mayor?' asked Socrates, looking to Ariadne. 'Why did you speak to him?'

'Yióryio took me to him,' I said.

'There is no chance of upsetting a Greek,' said Socrates, flicking his beads. 'Anyway, the school will not be ready for months.'

'It won't be ready?'

'Pano Anissari needs it for their Easter celebration.'

'But it's more than two months until Easter.'

'Then you should work here for two months.'

The villagers and the mayor wished to oblige. The Greek tradition of *filoxenía* – hospitality – made it difficult for them to refuse the request of any guest, even if they were unable to fulfil it. Yióryio had spent half a day driving me around Apokoronas when there was a suitable building 200 yards from my front door. In Greece the definite was an elastic notion and I needed to understand the different degrees of affirmation.

So I agreed to take the garage for a month. Or two.

On the drive back to the house Ariadne said, 'I want a list of questions from you.'

'Questions?'

'About construction and wing fabric. Have you calculated the centre of gravity?'

I'd never heard of centre of gravity.

'And what about the engine and propeller? Do you have them?'

'I only have a plan.'

'Then write out a list of questions.'

I thanked her for her help.

'Don't thank me. I do it because this aeroplane is a dream, and there are not enough dreams in the world any more.'

To begin I needed wood. According to the instructions the fuselage, struts and beams were to be cut from a single, twenty-foot length of straight-grained Douglas fir. Which grew in America and northern Europe. Manólis the village carpenter put his hand on my shoulder and assured me that he could help. He happened to have a piece in his *apotheki*. It was Douglas fir. Or as good as Douglas fir. But when I arrived on Monday morning to collect the timber it split in two.

'Lucky it happened to me,' he grinned, looking on the bright side, 'and not to you at fifty metres high.' He was pleased to have saved my life. 'I don't want you to fall out of the sky.'

'That's kind of you,' I said. 'But where will I find my wood now?'

'My brother is also a carpenter. In Heraklion. He will bring a new piece tomorrow. Or after tomorrow.'

'After tomorrow?'

'Before Friday for sure.'

That afternoon I drove to Heraklion to collect the long board and saved myself a week. I was learning to read between the lines, or at least to dissect the promises.

On his bench Manólis and I cut the timber into straight lengths, then planed them true, filling his workshop with the smell of pine trees.

'This wood is excellent,' I told him.

'You are a good man,' he said, 'and so is my brother.'

Next I needed extruded polystyrene to make the wing ribs. The Vrysses ironmonger sold door-size sheets, used for roof insulation, as well as shotgun shells, sheep shears and chain saw by the metre. I bought seven sheets and a five-litre pail of PVA white glue which was 'Suitable for dusty surfaces'. I also picked up a copper goat bell to hang from the Woodhopper's tail.

In the garage Katrin and I laid out the materials. She was excited by the novelty of the bare floor, by the volume of empty space, by our having achieved even this much. In a far corner of her mind, in a glimmering and unformed way, she was aware too of the symbolism of her – and our – performance that day. But the dust and concrete were much more real and she stamped on her fears.

Ulysses the village idiot perched himself on a stool in the garage as I sorted the cut lengths. I selected the finest piece, with the straightest grain and no knots, for the boom. I turned it in my hands, looking for flaws, orienting the wood. I tried to imagine its broomstick-slender length in flight, moving with the air, flexing up and down like a bird's tail, rather then from side to side like that of a fish. I marked along it the mounting points for anchors and spar brackets, rudder pulleys and aluminium down tube.

I proceeded with care, measuring two or three times before cutting the wood. As I worked Ulysses stared over my shoulder. I rested the boom on scrap timber when drilling to reduce back splinters. I made as few holes as possible to retain the timber's natural flexibility. It was against my nature to be so precise. In the past my carpentry had tended to be a blunt-pencil affair, a rushed muddle of ill-fitting offcuts botched together with speed and plastic wood. But then my life hadn't before depended on my work.

'Rub a candle along the sides of your saw,' Manólis advised me, 'so it will cut easier.'

I chose the two next best lengths of timber for the central wing spars. They too needed to flex on the vertical. These pieces had knots, and the grain was not straight, so I positioned the weaker parts of the wood away from points of stress, again flipping and flexing them as I worked.

'Left wing. Right wing,' I muttered. 'Left wing. Right wing.' Everything had to be made in pairs. The two wings needed to mirror each other.

Outside in the warm rays of the setting sun Katrin scribed the airfoils with the paper template, marking out twenty-two ribs on the polystyrene. The jigsaw kicked up plumes of white dust which clung to our clothes and the trees growing in the *platía*.

When dusk settled around the village we swept down the work surfaces and re-sorted my tools. Ulysses left his stool and wandered off into the night singing a tuneless song. A sweet-faced crone with grimy hands beckoned us to her doorway and offered us cinnamon biscuits. Beneath an apricot sky wives picked *hórta* for supper. Their husbands sat in the *kafeneion* where we stopped for a glass of wine.

'Is finished yet?' asked Yióryio.

'*Siga, siga*,' I said. Slowly slowly.

'By Saturday then?' asked Polystelios. He had tucked a purple periwinkle into his lapel. 'We could fly to Hania market on Saturday.'

'You'll never fly a Vespa,' said Socrates.

'Or ride your Aphrodite,' added Iánnis.

In the morning the garage doors had been pulled open for prying eyes. Ulysses took his usual place in the shade, trapped in a near horizon. Katrin and I started making the right wing, placing the timber struts across the floor, and found that the concrete wasn't level. It undulated along the walls and sloped towards the centre of the room. Using two spirit levels we chocked the struts with strips of cardboard and nub ends of polystyrene. Then screwed and glued into place the two diagonal compression beams for added strength.

As a result some of the joints were gappy. Too gappy. Which worried Katrin. I'd never been adept at cutting angles and relied on liberal applications of white glue. Now I tried to picture my gappy joints in the air. In the aviation classic *Stick and Rudder*, Wolfgang Langewiesche had written, 'Flying is done largely with one's imagination.' He considered aviation to be an art because 'we can't see the air, hence we often fail to hit it right, and hence so many of us break our necks'.

He hadn't mentioned the effect of gappy joints.

Around lunchtime Leftéri, five years and a few months old, tumbled into the garage. Yióryio's youngest child, chestnut-haired and stocky, was a baby Pan pushing a Tonka Toy bulldozer. He was the child who had recovered from meningitis and the experience had endowed him with brazen self-confidence.

Katrin and I were fitting the eleven curved ribs to the struts, adjusting angles with a T-square. Leftéri asked in a clear, low voice to see the aeroplane. When I pointed at the wooden frame he kicked away a chock. He had expected more progress. He then ignored my warning and like a boisterous goat drove the bulldozer into a stack of ribs.

I took him by the ear and banished him from the workshop. Leftéri screwed up his face, eyes frustrated now, and flew back as a fighter plane, with wings swept back, and machine-gunned the garage. We marched him home to his father who scooped him up into his arms, kissed him twice and gave us ice cream.

All afternoon we cemented the ribs onto the struts and by nightfall a skeletal wing had taken shape on the floor. We stepped around it with care, willing the cement to bond, not wanting to disturb the frame.

Manólis appeared at the door with a flask of *tsikoudiá* and two glasses. 'So you will fly sooner,' he said filling the glasses.

After the drink the wing looked level.

On the way home we stopped again at the *kafeneion*. It was impossible to walk past it without being dragged into conversation.

'*Ela! Ela!*' called Yióryio. Come in. Sit down. Tell us your news.

Papoos was shouting at the television. Kóstas beat Iánnis at cards. Ulysses gripped the seams of his trousers, pulled them above his shoes and exposed his ankles. We asked for a glass of wine. Polystelios hijacked the order and instead we were served more *tsikoudiá* . Again I tried to return his hospitality.

'You cannot buy me a drink,' he said, 'but I will sell you land.'

'My pockets are empty,' I told him.

'But you are a rich man. You own an aeroplane.'

'And soon you will build house here,' said Yióryio.

I said that I couldn't afford to build a house either.

'A small house, one room only, with a goat and some olive trees, then you live in the village for nothing.'

'Except for the cost of aviation fuel,' said Polystelios.

Sophia appeared with tin plates of tiny, pink shrimp *mezés*.

'Yióryio told me that I'm not the first to build an aeroplane in Anissari,' I said to Polystelios.

'There was another *pilótos*,' he replied, winking at Yióryio.

'There were many *pilótoi*,' embellished Papoos, turning away from the news to elaborate the story.

'But one in particular,' added Polystelios, 'a *xénos* like you.' A guest and foreigner.

'When?' I asked, doubting their facts.

'Before my grandfather's time,' he went on, holding up his shoulders and spreading his hands. 'He built an aeroplane and flew it. Why not? But we will tell you no more until you show us your machine.'

'Then come now,' I said to them.

Katrin and I led the tipsy procession out of the *kafeneion* and across the square. In the dark I found the lock and heaved open the doors. I snapped on the light, stepped into the frame, took hold of its two struts and lifted. The wing was sixteen feet long, three feet wide and weighed almost nothing. I raised it above my head and it didn't snap into pieces.

'*Kalá*,' called the drinkers. Good. Well done.

'Boom! Boom!' said Papoos, his voice lifting an octave as he pointed his cane at me. When I looked alarmed he explained, 'In the war I shot down aircraft like birds. German aircraft.'

Yióryio cuffed me on the shoulder and said, 'You are making a story.'

Manólis inspected the woodwork. Kóstas commented on the plans taped to the wall. The priest asked again, 'How much will it cost?'

Only Iánnis remained silent.

No one knew me in Anissari. I had no history here. But through our actions the villagers were coming to understand us. It felt like a new beginning.

Across the *platía* swollen wasps had begun to carry off the discarded prawn shells. 'Look,' said Polystelios, 'there you have true Cretan flying.'

The next morning I needed to squint to read the Wood-hopper's plans. It took us three days to make the left wing, repeating the pattern of construction, levelling the struts, attaching braces and squaring the ribs. The ribs had to be mounted with equal tension. Any small flaw at the wing's root would be magnified along its length. Which would mean that the two wings would not match. And the aircraft would not balance. And I'd fall out of the sky.

Again I took my time, not rushing the work, continuing to measure three times before making a cut. I sanded clean drill holes and countersunk brace screws. I used fret cramps and picked out slivers from my fingers with my Swiss Army knife's tweezers. The routine of construction was hardly stimulating but the labour soothed my mourning. I found it satisfying to handle the wood and to give substance to thought. Even after I hit my thumb with the hammer.

All morning Leftéri watched us from the door, his arms clasped behind his back, his hair plastered to his forehead by Sophia. He was quiet and well-behaved now and we invited him into the garage and sat him on a chair beside Ulysses. He did not want to be sent away again.

At noon Katrin flagged down a pick-up selling artichokes, their heads stacked up to its canvas canopy. We let the first bread van, with its broken exhaust and fumigated loaves, pass and hailed the second baker when he tootled through the village.

We broke for lunch at two o'clock to eat Yióryio's tomatoes, Polystelios' olives, fresh bread and sharp *graviéra* cheese. Our thighs and knees ached from kneeling over the frame. I lay on the ground for ten minutes, uncurling my spine.

After his siesta Leftéri returned with renewed confidence and a paper aeroplane. He spent the afternoon sailing it around the garage, a wing above wings. Every time I used a power tool he jumped forward to watch, a studious look on his unmarked face, and moved back as I finished. Then, when we were gluing the last ribs onto the wooden frame, he began to collect offcuts to make his own aircraft. He plucked dabs of excess glue off the floor to assemble an extraordinary contraption of cardboard and carved polystyrene.

'It will fly better than yours,' he assured me with Cretan certainty.

The next morning I gave him a model glider which I'd brought from home.

We watched the season change along the street, on our walk between house and garage. The wild white almond trees puffed into glorious flower for a week. Bees swarmed around Kóstas' cherry tree, then moved on to richer pickings after two days. Anemones and wisteria reached their prime. Katrin heard the first cuckoo.

We too changed, swapping jumpers for T-shirts, trousers for shorts. The sun roused the world back into life and we folded away the heavy counterpane and rolled the Calor gas heater into the back of the house.

Katrin and I lifted the wings onto their noses, strapped them back to back against a table and glued on the finger-thin trailing edge. The instructions called for forty-four 'cap strips' to line the polystyrene ribs. These thin, flexible, wooden strips would provide an anchor for the wing fabric.

Manólis cut them thicker than instructed. He knew better of course. And told me that he did. Twice. The extra thickness posed no problem along the flat underside of the wings. But on top the five-millimetre strips wouldn't bend around the curved contour of the leading edge.

'Just hammer a nail through them,' he suggested.

Instead we sanded the strips by hand until they were no more than a millimetre thick, then glued and taped them down, hoping that the springy wood would not rip away the crests of the ribs.

On each wing the pair of ribs closest to the Woodhopper's body needed to be 'boxed in' and shaped into a solid aerofoil. I cut the panels square, sanding and bevelling their edges, filling the gaps with slivers of polystyrene.

My shirt and shorts were snow-white with dust. My fingers were sticky from pressing the glue into crevices. My tools, which at the start had been organized in neat piles, began to clutter the surfaces and went missing under Leftéri's squadron of whimsical gliders. But I found the boxing satisfying for the skeletal wings – though still weeks away from being covered in fabric – started to look solid.

By the end of the second week the unseasonal heat became intense. It crackled off the earth, soaked into the skin. At lunchtime it stifled all sound, driving men and animals under cover. The mulberry trees, which days before had been pollarded stumps, sprouted two inches of summer green. Sheep took cover in their growing shade. The glue dried faster than I could apply it.

In the evenings we walked home through extraordinary cool streams of air which trickled off the hills and hot pools radiating from sun-baked walls. Watchdogs barked at the dusk, at each other, at bats wheeling in the twilight. Above the village the snow retreated into clefts and crevasses and the mountains lost some of their loftiness. They seemed less and less to be a part of a separate world. After supper and typing at my laptop I sat in the darkness of the house imagining the next steps of construction.

By the Saturday all the wings' elements – timber, polystyrene and some fibreglass – had been drawn together. Only the leading edge, the line by which they would meet the air, remained to be made.

I mounted long blocks of polystyrene along the wings' length, angled them with a hacksaw blade then shaped them with an electric sander. Katrin followed behind me, working by hand and eye, sanding, softening and contouring the hard lines of wood and polystyrene.

'It's female,' she said, stroking the smooth, tactile curve.

We laid the wings on the floor, end to end, and they stretched across the length of the garage: balanced, featherlight, poised like dandelion seeds on the brink of a journey. They no longer seemed lifeless. The inanimate had begun to take on an identity. I had the urge to pick them up, to hold one in each arm, to run down the street with the wind in my face.

I had read of an old custom of village boys carving wooden swallows, adorning them with spring flowers and walking from house to house singing songs. But Leftéri and the other children who visited the garage brought no wooden birds. They came on bicycles, or when the school bus arrived, brothers and sisters running across the *platía*, asking when and where the Woodhopper would fly.

'It is five times beautiful,' Leftéri told them, as his father had told him.

The children wanted me to be looping the loop by the coming weekend, as did their parents and grandparents. But they were also willing to wait, chasing each other around the garage, their feet falling close to the fragile wings, shrieking and, in the case of little Demetrius, pulling down his trousers.

Ulysses stood among them, pressing the silent transistor to his ear, talking nonsense. Once I asked Leftéri what had happened to him and he answered only, 'Bees.'

One of Leftéri's sisters paused from the game and pulled at my hand. 'But will you be able to fly it?' she asked me, catching her breath.

'I will try,' I told her.

'Then you will do it,' she replied. 'And I will be so pleased when you finish to see in my own village this thing made by you.'

7. How Ulysses Went Mad

There were ghosts in Anissari.

The *telónia* rose out of rocks to bewitch hunters. The worm-like Vrouchos seized travellers crossing the Elliniki Kamara bridge, unless they first spat into his eye. The cobbler Kathahanas, who had sold his soul to the devil for twelve ducats, sat every night at his graveside mending shoes. During the day he tramped the road under a load of leather skins. Yióryio claimed to have walked beside him.

'I smell the old cuckold all over valley,' he assured me.

Turkish tourists were haunted by a headless Kalives priest, executed during the 1821 Revolution for refusing to convert to Islam and impaled on a flagpole. The ghosts of two Egyptian mercenaries, thrown into a crevasse above the reservoir, snatched children who wandered near to their 'Nigger Hole'. Polystelios blamed them for the ruin of his flock.

'Bastard blacks stole another lamb,' he complained every other week.

Each spring in a grove beneath the village, by the hidden Turkish cemetery, dusty voices whispered through the olives. Ulysses' mother, superstitious Chryssoula, called them 'the big tongue' for their sirocco song had enticed her husband away from Anissari, never to be seen again.

'She's full of bullshit,' jeered Little Iánnis, spitting twice on the floor to be sure. 'Her old man just ran off.'

The villagers' ghosts – whether haunting winds or headless priests – were intimate acquaintances. Like the ancients' gods, their personalities were at once timeless and malleable. Their characteristics were mixed and neighbours often argued about individual attributes. Except for one spectre of spooks. All the villagers agreed that the fiendish mishaps of early January were the work of the *kallikántzaroi*.

According to legend the earth was supported by the branches of a great tree. All year long the satyr-like *kallikántzaroi* chopped away at its trunk so as to bring down the tree and destroy the planet. Then on 24 December with their job all but completed, these loutish creatures – with asses' ears and outsize genitals – ventured up to the surface of the earth to cause mischief: frying frogs, terrifying widows, deflating men's erections and – according to Yióryio – pooping in the flour. At Epiphany, 6 January, Orthodox priests, tired of the infernal antics, went from house to house, muttering incantations and scattering holy water. The *kallikántzaroi* were driven back down to their big-tree Underworld. But during their absence the tree had grown whole. The thick-witted beasts took up their axes and began again.

Year after year the pattern was repeated, as unchanging as the passing of the seasons, causing little permanent damage. Apart from the day that the *kallikántzaroi* drove Ulysses mad.

*

In the winter after the spring which had taken away his father, Ulysses led his mother's goat into the hills, as he did every day, except Sundays and feast days when he helped his grandfather to clean the church. He was a slow boy who stood apart from the others, not only because he didn't want to snare sparrows or spy on Aphrodite riding the manager of the olive oil co-operative. Unlike them, he wanted to see the world.

In Greece it had long been the custom for first-born boys to be named after their paternal grandfather, so every given name reached back through the generations. His father had told him that the first Ulysses had been a heroic traveller who had explored the seven seas before his death.

'Ulysses was persevering,' he remembered his father saying. 'Ulysses was *polyainos*.' Much praised.

His mother on the other hand told him only that Ulysses had spent twenty years trying to return home.

'He should have stayed put and looked after his family,' she said to her son.

Ulysses liked to sit alone on the high headland, gazing out to sea, considering how to become an adventurer and not be late for his evening meal. He had inherited the fondness to wander, even though it had bypassed his grandfather, a basket maker who never ventured further afield than the reed beds of the Armiros.

On that early January morning Ulysses pondered thoughts of home and abroad as he led his goat by a rope away from Anissari.

He knew his valley well, recalling the tale of each field and wall. He walked above Kalavaki beach which old man Kargamis had bought for a dozen sheep. He followed the secret path around the cliff where Kizilaki's mule had lost its footing. He passed the grazing land near Agia Ireni, the collapsed church with its spring-fed font encrusted with eight centuries of lime. He knew where to pick the aromatic *díctamos* for his mother and where the other boys dug for Saracen gold. He was forbidden only to visit the grove in Mezaria from which his father had vanished.

There were hazards along his daily route – the nocturnal *telónia*, the deep caves and Papoos' field of blue box beehives – and he avoided them, even though they were not as dangerous as crossing the sea on a raft or flying to the moon, a journey which he and his mother had watched one night projected across the whitewashed wall of the church.

'Maybe Papá has gone to the moon,' he had said on their walk home from the film and she beat him so hard that he fell onto the road and saw shooting stars in the sky.

She hit him most days, more often when he stepped on the chickens' eggs or soiled his trousers, but then she always took him in her arms and cried into his thick, raven hair that everything was all right. Which it wasn't, of course. In the *kafeneion* the men called him *o trellós sti hóra*, meaning 'the fool in the village'. He overheard them while feeding scraps to the goat at the back door.

As the mist retreated from the sun, Ulysses passed the mouth of the cavern in which the villagers had hidden from both Turks and Nazis. He climbed up through the scrub of wild almond, high, high up to Drapanokefala, the mastodon-backed mountain, to his headland and view of the sea. There, as usual, he sat all day with the goat on the barren stone hillside, tasting salt on his lips and wondering if his father had sailed a raft over the silver horizon, drawn by the siren sound of the wind.

'Ulysses endures much,' he remembered, confusing hero and mortal. 'Ulysses is persevering.'

The mythical Ulysses had sailed for home after the siege of

Troy. But within sight of his Cretan namesake's headland a gale had blown his ships off course and onto the lands of lotus-eaters and the Cyclops. The 'much-roaming' hero was cast adrift, tossed between Malta and Italy, lashed to a mast, his ears blocked with wax, weeping 'for his foiled journey home'.

'He sailed this way,' Ulysses thought to himself, looking to the west, not forgetting his lunch. He ate a potato and finished his water. He fell asleep under a lone fig, forgetting too that the shadows of fig trees were heavy and brought on bad dreams. The big-tongue wind haunted his sleep.

Hours later he awoke to feel the cool of evening. The day had passed away, tendrils of mist sliding inland over the valley. He was late and must hurry home to call in the hens.

But when Ulysses turned to look for the goat it was not there. He had released the animal to feed in the thistles and it had vanished. He stood, tying the *kérkos* rope like a long belt around his waist, thinking of his mother's anger.

'*Kyría*,' he called. '*Ela, kyría*.' Come, lady.

Sunken clouds black with rain descended onto Apokoronas. Ulysses had stayed too long on the eyrie, which dark mists now seemed to decapitate from the body of the earth. He called for the goat again, taking a step.

There was a rustling noise behind him.

He turned in relief, but the goat was not there. Nothing was behind him. He strode forward and again heard the sound.

Ulysses ran two steps backwards and the noise moved with him. He wheeled around and saw the smooth black rock on which he had slept away the afternoon. He knew that the *telónia* rose out of such rocks after nightfall.

But it was too early for them.

'*Kallikántzaroi*,' he said out loud, startled by the alarm in his voice.

Ulysses felt the heat of fear rise in him, fear of both the *kallikántzaroi* and of his mother. He started to run. The wicked spirits began to chase him. He dropped the *díctamos* which he had collected. He stumbled over rocks which cut his feet. He forgot about the far horizon and the goat, wandering among the thistles, and ran for home.

'Papá.'

Ulysses tumbled downwards to the bank of cloud, his arms windmilling. He ran through the grazing meadow, around the goat-rocks, into Papoos' hives without seeing them, knocking down two of the blue boxes. The startled bees rose up from the ground, flew alongside the *kallikántzaroi* and buzzed up the legs of his trousers. Ulysses cried out, hitching his trousers up against his crotch. The gesture trapped the bees against his balls, which began to sting in their own fear. He fell down again, hitting his jaw on the rocks, rose up to run on, pursued by swarms of demons.

Ulysses didn't pause in the five kilometres between Drapanokefala and Anissari, even after the bees rolled dead down his thighs. He wailed into the village and out its other side, like a wild wind, sucking the men out of the *kafeneion* and women away from their stoves. They too chased him, calling for the Virgin to protect him, but he couldn't hear them for he was truly insane now, frothing at the mouth, his big eyes torn wide open.

Ulysses led the shrieking string of villagers to Mezaria, to the fateful olive grove. He fell onto the earth and at last the *kallikántzaroi* left him alone. His mother, Chryssoula, wailed that 'the big tongue' had now destroyed her whole family. Others too recognized the mischief of the *kallikántzaroi*. Until, that is, Ulysses' grandfather untied the dragging *kérkos* rope from around his grandson's waist.

★

Ulysses did not recover. His mother continued to look after him, and to beat him, but she never again made him tend the goat or chickens. The 'incompetent mouth' spent his days walking away from the village in all weathers, not so much to escape it but as if seeking another traveller on the road. Anissari was both his cradle and a noose. Passing drivers always found him and brought him home. Half a dozen times during our stay I rescued him from places as far apart as Vamos and Fones, in winter his skin white with cold, then in summer darkened by the sun to the colour of polished walnuts.

In the *kafeneion* Ulysses perched on his stool, his ear pressed to the broken radio, his gums around a chocolate croissant. Or a sweet cream *bougátsa*. The villagers tolerated him, rather than locking away their shame as is more common in Greece. Among them, Ulysses would whisper to himself private, nonsensical words, a sound something between the wind and a buzz of static. Then he would check his braces with long, tapered fingers, hitch up his trousers against ghosts and set off again along the road.

8. In a Clearing Sky

As fly-blown *mechanés* putt-putted home from the fields, the Lefka Ori stretched their flanks to bask in the dipping afternoon sun. Long blue shadows fell across remote chasms and folds. Distant outcrops seemed close enough to touch. Every stone along even the furthest donkey path blazed as if polished by Apollo's light.

Polystelios too stretched out and asked Yióryio to pass him a hammer. 'Why build the aeroplane?' he asked me. 'Buy my *mechaní* instead and save yourself the trouble.'

Mechaní did not fly. They were fettered, motorized wheelbarrows with open bench seats and flatbed backs made from off-the-shelf Ford and Opel parts. As their name testified they were the first powered vehicles to arrive in many Cretan villages. *Mechaní* simply meant machine.

'How much are you selling it for?' I asked him.

'A good price. Say, a quarter of a million drachmas.'

'Is that all?'

'It's a bargain. Especially as it's thirty years old.'

Polystelios lay under his dishevelled Hercules, my tools scattered around him on the garage floor. He was trying to fix the clutch without removing the transmission. Which was like trying to eat an unshelled walnut. 'Do you want a ride on it?' he asked.

'Maybe tomorrow,' I replied.

'With its two wings it'll fly better than your aeroplane.'

'You have wings?'

'At home in the chicken shed. I'll put them on and we'll fly to Heraklion in one hour.'

'More like one week,' said Yióryio. Beside him stood his son Leftéri, a sober-faced little man today more interested in the adult world than childish play.

'Or maybe you could borrow an aeroplane?' Polystelios continued while whacking the flywheel casing. 'That too would be easier.'

'Or make up the story,' suggested Yióryio with a wink.

'Like your myth of an early pilot.'

'I say to my only God that it is very truth,' he assured me, hand on heart. Yióryio hated to be doubted, especially when he was at his most creative.

'That *pilótos* flew from Aptera,' shouted Polystelios, running with the fiction. 'Like Daedalus and Icarus.'

'You are for *panegyri*,' said Yióryio. Crazy. After the lunatics who were brought to medieval fairs – or *panegyri* – for miracle cures. 'They flew from Knossos.'

'I heard they jumped off the cliffs,' he insisted, getting it wrong again. 'Pass me the bigger hammer.'

Aptera, an ancient city a sparrow's flight west of Anissari, meant 'without wings'. It was named after the Sirens who, defeated by the Muses in a musical contest, had plucked off their wings and had thrown themselves into the sea.

'Sirens, Icarus, first *pilótos*, now you,' concluded Polystelios. '*Apístefto kai ómos alithinó.*' Unbelievable but true.

The prosaic mixed with the fabulous; past and present overlapped at every point. Yet I could imagine Sirens and Muses singing in the skies above Aptera more easily than an early twentieth-century Cretan aviator. Soon they'd have me believing that the winged youth Phosphorus, the star which shone in the dawn sky, lived with their cousin in Athens.

Polystelios hauled himself out from under the chassis. He had grown tired of hammering and begun instead to check the clutch pedal. As far as I understood it hadn't worked for a decade. Leftéri passed him my favourite screwdriver to prise off the release lever and detach the cable from the fork. 'It may need a new spring.'

My tools were covered in grease.

Polystelios' *mechaní* seemed to have grown out of the land, or at least in a dark corner of some decrepit lean-to. It had started life as a chugging Hercules, then sprouted Kóstas' cast-off bulldozer cabin and a tarpaulin trailer. Wheel hubs had been bolted to its sides, one each for coiling rope and sheep wire. He had welded onto its roof a chicken cage and attached a bracket for Aphrodite's walking frame, to be used on market days when they motored to Vrysses, gear box screaming.

'I'm not jumping off any cliff,' I told them.

'For sure no,' said Yióryio, enjoying himself. 'First *pilótos* too not fly so far from Anissari. He fly right here on street.'

Only the day before he'd told me that the first road hadn't reached the village until 1957.

'Or somewhere else for certain,' confirmed Polystelios as the pedal assembly and cable sprang apart in his hand.

In Crete there was always more than one answer to a question. At first I'd been unable to explain the constant contradictions. Even simple facts like the age of a church or the time of the next bus seemed to be matters of debate and opinion. I was beginning to suspect that an interesting story was prized above a prosaic truth. The explanation for this obliging invention seemed to lie in the influence of myth.

Myths were rambling, fluid stories, mixed up over millennia. Each listener took what he or she needed from them, to cope with life or to explain natural phenomena, making each myth their own. Individual versions became equally valid, and Greeks became used to hearing the same story told in different ways. This at least was my theory, and it went some way to explaining why it was so difficult to get a straight answer in Anissari.

In the same way at Aptera, as at many ancient sites, fragments of the past could be pieced together to make the whole that was Crete's history, a whole that differed depending on what remains were found: shards of Minoan pottery, massive Mycenaean defensive walls, Roman cisterns, abandoned Byzantine monasteries, crumbling Turkish fortresses and German machine-gun installations.

'So where would your pilot have found his wing fabric?' I asked them, trying to bring the conversation down to earth. To move on with the construction I needed thirty-five metres of strong, light material. 'Ariadne told me to look in Hania.'

'Rethimnon,' said Yióryio. 'For sure.'

'Lassíthi,' said Polystelios with equal confidence.

I asked them to explain, as Polystelios reassembled the clutch pedal.

'In Rethimnon port living sail makers,' Yióryio said. 'Old aeroplane was covered with sail.'

'On Lassíthi there were a thousand windmills,' said Polystelios. 'The *pilótos* used windmill fabric.'

Both suggestions seemed reasonable. As did Ariadne's

recommendation of Hania. I had three options, none of which might bear fruit. Which was typical.

'I'm going to Hania at the weekend,' I said. 'I could start looking there.'

'To meet Ariadne again?' asked Yióryio, putting his hand around Leftéri's shoulder.

'About an engine.'

'Rethimnon is better,' he assured me.

'Or Lassíthi,' suggested Polystelios, wrapping a cord around the *mechaní*'s starter. 'You can find material and see the other flying lady.'

'Ariadne is not only one in Crete,' said Yióryio.

With one pull the *mechaní* coughed into life. Polystelios hobbled up into the driver's seat and depressed the clutch. It screamed like a tortured rat rolling downhill in a steel drum.

'Better,' little Leftéri said.

Crete rides as if on the back of a bull. Two great continental plates collide beneath its high mountain ranges. The island lies almost equidistant from mainland Europe and Asia, midway between Athens and Libya, a few days' sailing from Egypt. The pivotal location made it a meeting point between East and West, tradition and innovation, myth and fact.

Zeus, the legendary father of gods and men, disguised himself as a white bull to abduct Europa, daughter of the king of ancient Phoenicia. He carried her on his back across the sea from Asia, made love to her beneath a Cretan plane tree and planted his seed in the continent that would bear her name. According to archaeologists the first Cretans also crossed the eastern Mediterranean, settling near Knossos eight thousand years ago. They built flat-roofed mud-brick houses reminiscent of Pharaonic Egypt, cultured the olive and laid the fragile foundations of Europe's first civilization. Their rituals and myths were adapted to their new world, grounding their fears and imagination in the landscape. They created a world charged with divinity, worshipping the bull who made the earth quake, idolizing the fertile, feminine earth goddess.

Katrin and I drove east along their ancient north shore, between the sea and the mountains. Rocky hillsides wore yellow cloaks of broom and gorse. On the outskirts of Heraklion we passed Knossos, the palace of King Minos, first-born son of Zeus and Europa, whose name had been given to the Minoans. His palace was the tangible truth underlying the West's oldest stories. From here, according to Thucydides, Minos ruled the waves and rid the seas of pirates. From his storerooms exquisite pottery and woven wool textiles were exported to the Peloponnese and Egypt, as depicted in tombs from the twelfth dynasty. Minoan metal work and olive oil went to Sidon, Tyre and Troy. In the *Iliad* Homer described Crete as a rich and populous land of one hundred 'cities'. Athens, if it existed at all at the time, was a village.

According to legend, Minos' wife, 'lewd and luxurious' Pasiphaë, fell in love with a bull and from their union bore the Minotaur. This allegorical beast, half-man, half-bull, was hidden – along with Minos' shame – under the palace in an endless labyrinth built by Daedalus. Every nine years a tribute of young Athenians was sacrificed to the monster, as parts of the mainland and the Cyclades were then subject to Crete. Theseus, son of the King of Athens, volunteered to be one of the seven victims. Ariadne, Minos' daughter, lost her heart to him and, with the help of Daedalus, gave him a ball of thread to unwind as he entered the labyrinth. Theseus killed the Minotaur – Ariadne's half-brother – retraced his steps and escaped with Ariadne, only to abandon her on Naxos.

The betrayals foreshadowed the end of the Minoans.

In his fury Minos imprisoned Daedalus and Icarus in the

labyrinth, from where they escaped on wings. Herodotus recorded that an expedition of 'close-shaven' Cretans pursued Daedalus to Sicily. The great inventor eluded them but Minos perished when his fleet was wrecked on the Italian coast.

The eruption of nearby Santorini, the second largest volcanic blast in the last ten thousand years, caused tidal waves and climate changes which devastated the Minoan civilization around 1450 BC. The ravaged island fell to Greek invaders. The survivors – now interbred with Mycenaeans – took flight to the mountains and across the sea to Cyprus, Rhodes and – like Ariadne – Naxos. Greece plunged into a dark age of fear and violence, illiterate and without written records, from which the greatest legacy would be the art of oral improvisation.

Beyond Gournes we turned away from the coast, weaving over the passes through which the Minoans had escaped from the Dorians. The dry white pasture land fell away behind us. Along the roadside goats scratched themselves against new crash barriers and shepherds in trim black turbans watched their flocks from the shade of blue European Union works signs.

We pushed up the giddy, rutted road and into the Dichte massif. Its bare, intractable flanks filled our vision, forcing us to sit back in our seats, and all but blocking out the sky. Mount Ida reared up out of the sea, a Cretan Everest with a wasps' nest of lesser peaks around its waist. We paused for breath in Krási, a village of venerable plane trees and thick-trunked eucalyptus. Beehives swarmed around the church. Winter wood was stacked beside white houses. Chrysanthemums grew in olive oil tins. In places dregs of stinking, purple mash lay on the roadside, the lees of grapes and stalks discarded from last autumn's *raki* stills. In the taverna a pie-eyed priest told us,

'I made fifty kilos to get me through the winter. Stay in Krási and we will drink it together.'

In Crete, journeys must be measured in time not distance. It was less than fifteen miles from the coast to the plateau yet the drive took us two hours, and not only because of the second glass of *tsikoudiá*. In places the mountains so separated north from south that some hill villagers, though living within a dozen miles of the sea, had never eaten fish or seen an octopus until the first road was built.

At the crest we spied volcanic Santorini, seventy miles to the north, and caught a signal from a Libyan radio station, Arabic voices at the edge of Europe. We saw the openness of the northern shore and the narrowness of the island's wasp-waisted length. Then we tipped over the rampart of mountains and fell into a hidden world.

Lassíthi was a sunken patchwork of green fields ringed by serrated cliffs. Around its rim were eighteen terracotta-roofed villages, each within sight of the others, making intimate the broad, flat plain. We circled it, the highest occupied community in Greece, driving between plots of cereals and orchards of apple and quince. In summer packs of tour buses ascended from coastal hotels to jostle on the narrow streets with cobblers' vans and carts of potatoes. Village school children dressed like ersatz travellers, carrying kaleidoscopic mini backpacks and shouting '*Bonjour*' at cyclists drinking glucose drinks with their *dolmádes*. But out of season Lassíthi retained a cool air of mountain fastness: alert, self-contained and free of the eastern somnolence of the coast.

In Psihró – which means 'ice-cold' – a shrill matriarch tried to sell us a leaping bull T-shirt. Above her stall rose barren Mount Karfí, site of one of the defeated Minoans' last, grim enclaves. Behind it was their most sacred site. We followed a wooded trail to the dark entrance of the Dhiktean cave, then descended the damp steps into the body of the earth, between the ancient bones of stalactites and stalagmites, slipping through a narrow opening beside a pool. For thousands of years men believed that Zeus had been born in this rosy-pink chamber. Minoans and Greeks had venerated the cave, which was also the place where Minos was said to have received his Law Code from his father. At the start of the last century archaeologists found hundreds of votive offerings to the bull and the Earth Mother: bronze figurines, knives and miniature double axes. I too threw a coin into the birth chamber, as so many others had done before me, to mark my passing through a centre of their world.

I felt safe in the womb of the earth. Like the plain beyond its lip, it was contained by a ring of rock, as Crete itself was embraced by the Mediterreanean. And that ancient sea was in turn contained by the shores of Asia, Europe and Africa.

Thirty years ago the white sails of the Lassíthi windmills turned in the breeze, pumping water from cisterns to fields. On the plain's five-mile-wide expanse once stood the greatest concentration of windmills in the world. As many as ten thousand according to some reports. Now only a handful remained, most of them converted into gift shop billboards. The rest were rusty spindles of metal, tumbledown and superseded by petrol-driven pumps. But I reasoned that even the very few which remained, fluttering like butterflies in the cool spring air, needed someone to maintain them – and their sails.

We waded away from the raised ring road through a surf of flowers; daisies, asphodels and wild white irises. In places the rich, tilled soil looked so soft that Katrin wanted to lie down on it and sleep. Across the patchwork plain we could see a pair of working windmills and we made our way towards them. A man clung to one of the towers, silhouetted against the sun and waving his arms in slow, broad movements as if trying to levitate himself. As we drew near I realized that he was winding long lengths of heavy wire around the sail's axle to stop it from rocking off its rusty mountings.

'*Kaliméra*,' I called.

'*Guten Tag!*' he replied and slid down to earth. He wore a beard and a ponytail and in his Oshkosh looked like a denim hand-me-down from the sixties. '*Willkommen*,' he said reaching out a great callused hand. '*Kann ich Ihnen helfen?*' How can I help you?

'We're looking for the man who repairs the sails,' I said.

'No one repairs them any more,' said Michalis, slipping into English. 'They're too old.'

Michalis wore a leather tool pouch on his webbing belt with a drill holster. He was about forty, had Aegean blue eyes and smelt of cigarette smoke. I explained that I needed sail cloth.

'For a boat?' he asked.

'For an aeroplane.'

'Are you the *pilótos*?' he cried, grabbing my hand and pumping it. '*Kalós orísate*. I heard about you from my neighbour.'

Even though he had no sail cloth Michalis invited us for coffee. His house was across the fields in Tzermiádho, the largest village on the plain's northern edge. In an orderly showroom beyond the Kronia taverna were blue coils of plastic pipe and half a dozen dependable, uninspiring, electric water pumps. Callipers, snips and hand-cut rifflers were arranged in precise rows on the wall. Metal sheet was stacked according to grade beneath the table. His bed was behind the pedestal drill.

'My grandfather helped to build the first windmills,' he told us.

In the 1920s the windmills had replaced the dipping wells which had irrigated the plain since Venetian times.

'When I was a boy there were thousands of them, as necessary to Lassíthi as a copper pot to a bride.'

His father had continued the tradition, building and maintaining them until he was knocked off a tower by a cockeyed sail and drowned in the cistern.

'But now they are old news,' he said, dismissing the past with a flick of his hand. 'They're too much work, like copper pots.'

Modern Greek brides preferred matching sets of Teflon-coated saucepans, not copper ones which needed cleaning with damp straw so they gleamed like a mirror.

'Not so long ago the last tinker in Greece appeared on television begging for an apprentice, offering to give away his equipment. No one wanted to learn what he knew. Can you blame them? His tools went into a museum.'

Michalis was the local distributor of Deluge water pumps. He was also a kind of post-modern Hephaestus, the ancient god of metal work and contriver of mechanical devices. Except his inventiveness had been supplanted by instruction manuals. He couldn't understand why I hadn't bought an easy-to-assemble kit aeroplane.

'Because most of us have lost an understanding of the workings of things,' I explained. It was a part of my need to reach back to beginnings, to make something my own.

'Flick on a switch, it works,' he replied. 'If it doesn't, throw it out.'

Mythical Hephaestus had been a goldsmith and an armourer. He had made Achilles' shield and four magic fountains which spurted milk, wine, perfume and, depending on the season, hot or cold water. After seducing Europa, Zeus had given her three gifts: a dog to bring her food, a bow and arrow for her personal safety and the giant Talus, guardian of the new community. Made by Hephaestus, Talus was a brass robot who circled Crete three times a day, throwing rocks at enemies or hugging them to death against his red-hot chest. Like the Minoans, Talus was defeated by Greeks. The giant's life depended on a vein of blood which was sealed with a bronze nail. Medea, sorceress and wife of Jason, told Talus that if he took out the nail he would become immortal. He believed her, wrenched it out and, as his Cretan lifeblood spilt away, the unconquered island fell to the Argonauts.

'Is that aluminium?' I asked Michalis, pointing at a bin.

The dozen lengths of metal had caught my eye. I needed twenty-six feet of 1⅛-inch diameter tube to make the Woodhopper's open air cockpit and landing gear.

'It's for irrigating a pig house,' he said.

The tubes felt light and not particularly solid but Michalis assured me that they were the best quality. Especially when I asked to buy them.

We spent the afternoon cutting lengths, deburring edges, setting angles and drilling holes. It was a very Greek day. We had set out in search of one thing and found something totally different which might or might not be worthwhile.

'*Einai Ellada,*' said Michalis. 'This is Greece.' Life is in the lap of the gods.

As we loaded the tubes into the car he pumped my hand again. Katrin asked him where to find Lassíthi's 'flying lady'.

'The Chained Lady?' he replied.

'We were told that she had wings.'

'They are one and the same.'

Beneath a ridge of ruined stone windmills we found a concrete convent set amongst mulberry trees. Panayía Kardhiótissa, meaning 'She who is in your heart', appeared at first to be an unremarkable *souvláki* of mismatched buildings. But inside it revealed itself as one of the island's most important places of worship. Exhausted, wide-eyed saints graced the fourteenth-century Byzantine frescos. On the iconostatis was the 'wonder-working' flying lady.

Crete had drifted through the Classical and Hellenistic periods on the periphery of the Greek world. It had fallen under the Romans and then, after the division of their empire, to Byzantium with its absolute rituals and calculating ways. More than seventy fine Christian basilicas had been built across the island during the period as well as thousands of tiny, exquisite medieval chapels.

After Constantinople was sacked by the Crusaders, the island had become a Venetian colony. Over the next 465 years the Cretans rose up twenty-seven times against the occupier, only succeeding in 1669 in replacing the Most Serene Republic with the Ottomans. But the Turks wreaked havoc on the island, as they still do in every Greek mind, and the islanders didn't gain independence and union with Greece until the start of the twentieth century.

'The Turks stole the icon,' explained Michalis, 'and took her to Constantinople but she flew back to us across the sea. So they stole her once more, and again she flew home. After their third attempt she was chained to a marble pillar in Turkey. The next morning both she and the pillar were gone. She had carried it home with the chain still around her.'

We had seen the pillar behind a fence in the courtyard.

'It says here,' read Katrin, 'that the original icon of the Virgin was taken in 1498 to Rome, not Constantinople. And that this one is a copy.'

'They are both true stories,' said Michalis. Of course.

9. The Barber

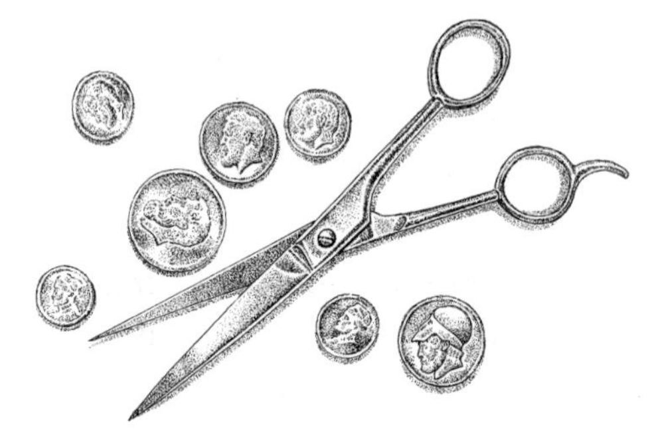

'I will cut you,' said Little Iánnis.

'Pardon?'

I half-expected him to pull out a vine-pruning knife and slice my throat.

'Your hair is too long, *kalamarádes*,' he went on. He meant pen-pusher, from *kalamári*, the ink-carrying squid. Of all the villagers he alone didn't call me *pilóte*. He picked at my ragged locks. 'Come eat with me. I will cut you and tell my story.'

'This is my village,' he told us in German, his voice hoarse as if he'd been shouting at a football match. Silver-framed tinted spectacles hid bruised eyes. 'My mother lives here. My home is here.'

We had followed him to the sparse end of Anissari. Spindles of rusted metal poked up from the corners of his bare concrete bungalow, waiting for the money to build the upper storey. Between bags of hardened cement Iánnis' mother grew a few vegetables, tended her chickens and nurtured a pocket-handkerchief garden of geraniums.

'But I was born there,' he said, jabbing his dull scissors beyond the open door, across the valley, at a pristine white house with closed blue shutters. 'There,' he repeated, 'Germans now own.'

I sat in an upright chair in the unfinished kitchen, my hair falling in casual tufts onto the floor. I'd wondered how a disabled barber with only one working arm managed to cut hair. The answer was slowly.

Half a dozen fitted units stretched across one wall. The other was lined with threadbare sofas. In the middle of the space was an archaic wood-burning stove. There was nothing else in the room. The living room and bedrooms remained to be built for the wife and children who would never come to visit. He and his mother slept on the sofas.

'If this is the last time that I live, I want to be here in my village,' he said, slicing around my ear. 'It is my life here.'

'But you stay in Athens.'

'In Athens I forget my name. You understand? It is a desert. But I must live away because of the accident of my life. I show you.' He put down his scissors and pulled up the sleeve of the paralysed arm. A lattice of deep scars ran up over his shoulder to a long, amber wound.

'I am the cripple of Anissari,' he said, squeezing the words out of his larynx. 'The barber. The German.' He looked raw, unkempt, corrosive, apart from his hair which appeared neat even after hours of drinking. 'I can never live in my village.'

Forty years ago a German entrepreneur came to Apokoronas looking for property. He saw the pretty family house in Anissari and made an offer. His timing was lucky. Iánnis' father, also a barber, had been killed the month before in an explosion at the Asprosikia quarry. To the bereaved family the offer seemed heaven-sent. A gift from Saint Antonis himself. But it was also a trap. Iánnis' mother accepted it and with the money set her son up in business.

His barber shop was long and narrow, only an arm's width wide, with one chair, one mirror and photographs of miniskirted calendar girls in woven hats. The walls were covered with psychedelic flock wallpaper, which he had bought in Hania to attract a young clientele. There was a canary, a black and white television and a single cold-water tap. Every morning he filled a cork-stopped thermos with boiling water at the *kafeneion*.

In his black trousers and white smock, Iánnis prospered for, unlike other barbers, he didn't talk while working, which was an innovation. He would shave a man with a straight blade, working up a thick lather, adjusting the sideburns with a finger-flick of moisture, all without uttering a word. His barber shop became a refuge in the Anissaris, if not Apokoronas. He earned seven drachmas per cut.

During that first summer, groups of students from Munich rented the old family house, which had been modernized by the new German owner. In those days there were few tourists on Crete – the first charter flight didn't arrive at Hania airport until 1974. Foreigners, especially blonde, blue-eyed Germans, were as rare as quiet barbers and the girls reminded Iánnis of the picture-perfect models pinned to his psychedelic walls.

He needed no pretence to talk to them. After all, they were sleeping in his old home, or at least the shed where his father had stored the animal fodder. He called on them, brought them *paidhákia* lamb chops cooked by his mother and showed them the ruins of Agia Ireni. He took them to Kefalas, the secret swimming cove beyond Paleloni, and it was there, beside the silky water enclosed by a rocky rim, that he first made love. To two of them. They gave him their young, tanned bodies with no notion that the gentle act would screw up his life.

In those days Iánnis had been better looking, with luxurious eyelashes and small, even teeth. His eyes had been as clear as the cool waters of May. His open-handed gestures made him appear tall. He wore blue jeans and a clean white shirt, ironed by his mother.

Iánnis became Apokoronas' first *kamáki* boy, meaning one who fishes with a trident spear. At the start of every fortnight he called by his old house, on its new lodgers, introducing himself and them to the charms of Crete.

'Are you famous in your country?' he asked, if the women seemed willing to be speared. 'No? Then I will make you famous in mine.'

Over the summer he slept with half a dozen Berliners, the two Bavarians and a dental assistant from Bayreuth. He made love by the sea, in his uncle's boat, in the back of a Mini-moke with an Austrian who wailed so loudly that his mother called the police. Iánnis caught a glimpse of a new world and liked what he saw.

Then, at the end of holidays, the tourists went away, taking his dreams with them.

That winter in the barber shop the only skin he touched belonged to his regulars, rough stubble in need of shaving, Nivea and talc. As he worked he resolved to move to Germany. He would open a salon in the wealthy land of sex-hungry blondes and find himself a wife. In February he closed the barber shop, took his money from the sale of the family house and moved to Frankfurt. He rented a small shop in the suburbs and opened his salon. It was a failure.

In Germany no German woman wanted either him or his quick-clipping hands.

'After that how could I come home?' he said to me. 'You want that I ask my mother for the money just to buy one coffee? That I cannot do. Thirty years ago there was no work in Anissari.'

In the sixties hundreds of thousands of young Greek labourers were contracted to work as *Gastarbeiter* in Germany. Their earnings were sent home to feed the family, to educate siblings and to finance the concrete house to which, months or years later, they would return to marry a village virgin. It was the eager dream which in reality delivered only toil, sweatshop wages and loneliness.

Iánnis had no formal schooling, like most *Gastarbeiter.* But he did have a godfather – a *synteknos* – who found him a job as a boilerman in Bremerhaven after the failure of the salon. He met a girl there, not an Aryan seductress but a hard-working, tawny-eyed chambermaid from Corfu, married her and had two children. He found the foundry work arduous and asked to be transferred to the company laundry. He was first to be fired during a round of cutbacks. The godfather then paid for him to work on a German ship.

'For five years I was on refrigerator ships,' said Iánnis, as he clawed at my hair. 'One day I went from the deck – where it was thirty degrees of sun – to fix the freezer. I was high up and suddenly all went black and I fell seven metres. Now this arm does not work and I am half; half a man, working as half a barber.'

After the accident Iánnis returned to Greece but, haunted by failure, came no closer to home than Athens, only stealing back to Anissari from time to time to see his mother in the half-built concrete house. He still paid rent on his wife's flat in Germany.

'My wife, *macht nichts*,' he said with a wheezy laugh. She doesn't matter. 'There are lots of women on Crete. Lots of Russian women.' Meaning prostitutes.

My haircut was finished. Iánnis topped up his ouzo. Katrin and I were drinking wine. His mother came in from outside with a dinner of steamed snails and kid goat. Iánnis scooped the snails from their shells and fed them to us. He had no plates, only the serving platter.

'The life is not easy but we make it so civil here,' he said. 'It is my last hope that I can come back and stay in Anissari. It is something that I have in my heart.'

Before Iánnis had moved to Germany he had owned twenty animals – goats and lambs 'for eating in the grass'. Anissari had been poorer then and less well educated. But the villagers may well have been a happier people, the rhythm of life unchanged for hundreds – even thousands – of years. Existence was a matter of routine. Men did as their fathers had done. There was a sense of sharing, working together in the olive groves or drilling for water, picking vegetables in the fields, standing shoulder to shoulder, touching, confronting common problems and curiosities: the lack of rain, death and falling crop prices.

Now his widowed mother kept only a few chickens for herself and watched television soaps.

'Tell me now, how do you feel for my country?' he asked, mercurial, unstable, shouting as he became drunker. 'In Crete all the persons is one person,' he continued without waiting for a response. 'Five persons is one, and one is five. All the people love this country, and I hope you do too.'

I assured him that we did.

'Do you know what I spend in a year buying drinks in *kafeneion*?' he asked. 'Half a million drachmas! Every visit a thousand here, two thousand there. And for what? To go back to Athens and fuck kill my life.'

'Little' Iánnis was the village's outsider, to be bullied and used because he had spent too long abroad and come home without a wife.

He wailed, 'We all come out of the same hole – Greeks, English, Germans – and we all die. Only this is certain.'

Suddenly he stood up, threw open a fitted closet and grabbed a double-barrelled shotgun. He loaded it with two cartridges and, standing in the middle of the living room, fired it with one hand out of the open double doors towards his old house. The deafening echo rang cold in the concrete shell.

'Now you,' he shouted at Katrin in a sudden rush of fury. 'You shoot the bitches.' And Katrin, astonished at herself, fired the shotgun.

Iánnis reached to shake my hand, hooking his eyes onto mine. His skin was rough.

'You have found my village. I have lost it.' He had about him a look of ruin. 'Write in your book all you want about Anissari,' he said. 'But if you write something stupid I will kill you. The next time you are dead.'

His mother slipped in to clear away the dishes.

10. Greeks Bearing Gifts

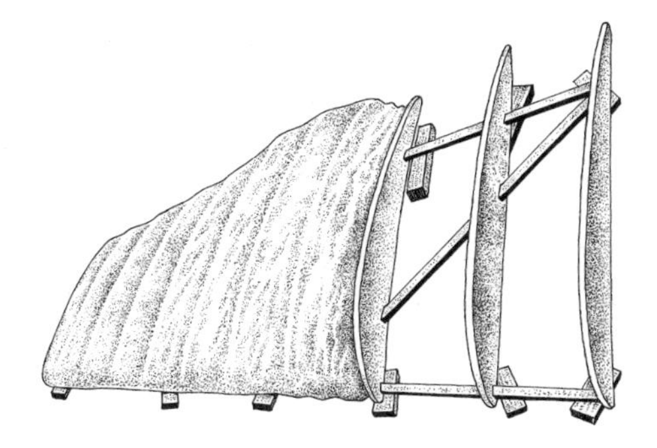

'Stop in Vrysses for everything,' read the sign above the town's mini-market. We did, and found that the storekeeper was Iánnis' niece. She was also a dressmaker and the local lonely hearts columnist. Her twice-divorced son Pascali supplied her with both job lots of imported cloth and soul-stirring sagas. I didn't want to hear his story too. The dying embers of the barber's anger had reminded me of my own. Defiance willed the aeroplane forward. Pascali ordered me thirty-five metres of an unshrunk material called Ceconite. It seemed ideal, especially when we unrolled a bolt on which was printed 'Suitable for FAA Part 103 usage'.

I allowed four days to cover the wings. It took us over a week. And I aged a month with the frustration.

On the first morning we laid out the wings' skeletons. They needed to be sanded then cleaned before attaching the fabric. We sponged off the saw and polystyrene dust as if washing down a corpse. With a clean cloth we wiped down the ribs, feeling the feminine contours and bone-like spine of struts. Every hour I knocked at a neighbour's door with our empty bucket, which he filled and refilled through the day, each time plying me with a glass of wine and a taste of his wife's *mizíthra*.

On the second day we sealed the wood and polystyrene with three coats of diluted white PVA. I knew and liked the glue, having dribbled it across most of my parents' carpets. As a boy I had used it to build model aircraft and insect-collecting boxes (with carrying handle and small glass window so captives could enjoy the view). Once I had made a bird house out of *The Boy Mechanic*, with veranda-like seed tray and gluey picket fence feeder. Whenever a bird landed on the perch a micro-switch rang a bell in our kitchen. My mother had found it to be an entertaining innovation, though after two mornings of being woken at dawn I was persuaded to disconnect the wiring.

As I brushed on the first PVA coat five-year-old Leftéri flew his tricycle across the road whistling to himself. Or, at least, trying to whistle.

'Is that your special song?' I asked him.

'It's birds singing, *pilóte*,' he replied.

I finished the last coat at dusk, painting by memory in the shadow of the single light bulb.

The next morning we began to cover the right wing, hoping to complete the job in a day. I unfurled the gossamer-thin fabric, dressing the bare skeleton in a white shroud. We teased the material into position, trimming it with pinking shears, and glued it to the wing rib by rib, watching for cement spills. The method was no different from that conceived by Leonardo five hundred years before for his *ornitottero*.

The Ceconite gave the wing a look of completeness. It

no longer appeared a hodge podge of cap strips, screws and timber, rather, it became an entity unto itself. As we eased the fabric taut it was transformed from a rumpled bridal bed to a ship's sail; from a sleek parcel, wrapped and hiding secrets, to a white cocoon enclosing a wish.

In contrast the left wing, waiting on the floor, looked naked.

For all its beauty I found the work tedious – stretching, straightening, applying glue, waiting for it to dry – and tried to maintain my concentration by imagining the wing cutting through space with me suspended beneath it. It wasn't that I preferred the flying machine to remain an illusion, rather that the prosaic nuts and bolts, minutes and days, seemed so far removed from the weightless flight of my dreams.

'That's good enough,' I said, slapping extra glue onto the root panel, obsessed by a sense of urgency.

'We'll do it properly,' said Katrin, proceeding with care, preferring to keep me alive. 'Or not at all.'

That third day we only finished one side of one wing. Each morning I set us a goal – to shape the leading edge or to seal the pear-drop patches – and come evening we were usually lagging behind. A conspiracy of hospitality, blunt pliers and slow-setting glue frustrated my plans. My reach may have exceeded my grasp . . .

I acknowledged that I was a guest on Crete. I respected the island and its traditions, even though the Greek sense of time was at odds with my own. Yióryio had laughed at me, or the *pilótos*, for importing tools and fabric, yet he was still to bring me a seat belt promised more than a month earlier, and I hated him for thwarting my schedule. If I didn't finish the flying machine by May the summer winds would delay my flight for months, stranding me in limbo with my demons. I had no doubt that Yióryio would keep his word, in his own time. Cretans weren't lazy or inactive, they simply had more important priorities than following others' plans.

It took us another day to attach the fabric to the wing tips and leading edge, teasing it around the curved panels, smoothing away each minute bubble with palms and fingertips.

'I think we'll need that help soon,' said Katrin, catching her breath as we flipped the second wing onto its trailing edge.

As the work became heavier I was pushing her harder. I could no longer lift a wing by myself and their span made the wings awkward to handle. Ahead lay the even weightier metal work, wire-making and fuselage assembly.

'I'll call Ariadne,' I said, impatient and heartsick.

My absent engineer had promised to come days earlier. She had also offered to find us a helper.

'Twelve horsepower?' scoffed Apostoli. 'That's the strength of my olive strimmer.'

A Greek god stood before us: tall and taut with a neat face, sharp black eyes and pursed, overhanging top lip. He radiated luminous confidence, which made him appear older than his twenty-one years. His English was precise and fluent. His clothes were crisp and his hair groomed, in contrast with my paint-splattered sweatshirt and glue-covered hands. Nothing in his features was loose, lazy or extravagant, apart from rather comical bat ears which seemed ready to fly off his head.

'I've seen him on a Grecian urn somewhere,' Katrin whispered to me.

'The wings are too big,' Apostoli said with disdain. He pushed back his dark glasses, drained a can of Coke and cast a critical eye over the Woodhopper's plans. He was Apollo. Or Ajax. I half-expected him to flourish a thunderbolt to strike down a passing Trojan. 'Why have you not installed a hook on the boom to determine the centre of gravity?'

Little Leftéri stared in wonder at the deity. He picked up the discarded Coca-Cola can as if it might be a divine relic, then thought better of such reverence and threw it out of the door at the goat tethered in the next field.

'Apostoli has time,' said Ariadne, supervising us from the door. 'Lots of spare time.'

The god was her son.

My short engineer and her bat-eared Olympian had driven out to Anissari in separate cars, his supercharged Fiat chariot and her clapped-out no-name estate. She wanted me to take him on as our assistant. To bestow a heavenly blessing on the project. And to distract him from drinking coffee and chasing girls.

'My greatest wish is to be a pilot,' he told us.

I imagined him as an Ionic figure winging across the far hills. Or an airborne Hercules, wrestling with a GPS Navigator if not with the Lion of Nemea.

'Flying is freedom,' he said. 'You can see the world from the top side. You can measure the space that you have down on earth.'

'He's just a kid,' said Katrin in my ear.

'It would be a privilege for me to help you,' he said.

I explained that I could not pay him. Buying the bits for the Woodhopper had absorbed most of my small inheritance.

'Thank you but I want no money,' he stated. 'It is enough for me to give wings to a dream. When do you want to fly?'

'Before the summer winds.'

'That is enough time,' he pronounced with the certainty of the Delphic oracle. 'This aircraft is very simple.'

'There's also the question of where to fly,' said Katrin.

'Maleme would be best for you,' he said. 'It is a big airfield and if you miss it you can land on the beach or in the sea.'

'Maleme?' I said. 'But Maleme is an air force base.'

Maleme was the island's original airfield. Hurricanes and Messerschmitts had landed on its runways. Above it the white canopies of history's first airborne invasion had opened. Its loss – along with many British, Australian and New Zealand lives – was crucial to Germany's victory in the Battle of Crete during the Second World War. It did not strike me as an auspicious place from which to fly.

'I'd never get permission from the military . . .'

'The commanding officer is a friend,' interrupted Ariadne.

'It is often best to speak to officials to arrange unofficial matters,' said Apostoli.

'Of course we cannot tell him too much because the air force has certain rules,' she admitted.

'About amateur pilots flying unlicensed aircraft from a restricted airstrip?' I asked.

'A single flight at the weekend will be no problem,' Apostoli said. 'During the week the airfield is used for target practice.'

'Great,' said Katrin.

'We have a saying, "If nobody accuses then there is no court."' Meaning laws can be broken as long as you're not caught.

'Tomorrow morning we start,' said Apostoli, looking at his watch. 'But tonight is Friday and I am meeting my girlfriend for coffee.'

Drinking coffee was a euphemism for a night out. A coffee might last an hour, or until dawn. On cue his Nokia rang with the theme tune from *Top Gun*. He answered the call. The conversation was short.

'You know, I am twenty-one years old and all my nights are crazy nights,' he said as he slipped the mobile back into his bum bag. 'She misses me already so I must leave.'

*

'I passed through hell to keep that kid,' said Ariadne.

In the *kafeneion* the men were playing cards and, rather than face their questions, we moved a table outside under the olives.

Ariadne wanted to talk of her 'interesting days'.

At the age of twenty-four, home from her first year at Boeing, she sat her mother at the kitchen table and said, 'Take your heart tablets now, there's something I need to tell you.'

'I don't need my tablets. I'm fine.'

'Take them,' she insisted, then told her that she was pregnant. 'Not by Yióryio. Not even by a Greek. By a foreigner. A man who may have a dozen families for all you know.'

Her mother shouted at her, cursed her to the devil and drove her out of the house under a barrage of matching crockery. Ariadne arranged the wedding herself, with her fiancé's help when he flew in from the States. Her mother would have wanted a white wedding for her only daughter – even if she was a foreigner's whore – and Ariadne obliged. She met the local bishop when summoned to see him.

'My child,' he said to her, laying open the fiancé's passport on the neat desk, 'we will need letters from all these countries: England, France, Switzerland and the United States.' All the countries for which there was a tourist entry stamp. 'To confirm that your fiancé does not have a wife and children there.'

'Why do you need to do this?' asked Ariadne.

'I am trying to protect you,' said the bishop.

Ariadne, who only tiptoed through the theatre of Orthodoxy to please her mother, saw black and red spots before her eyes. 'Listen,' she said, 'I didn't use protection and now I'm pregnant.'

'My child, the foreigner made you pregnant?' said the

bishop, all syrup and sympathy. 'I understand now. Maybe it would be best to delay the wedding. There are arrangements that can be made to help to relieve you of any pressure to make a decision.'

'I want to marry him,' she said, her anger flaring, 'and if you stop me I will tell the newspaper that you are advising women to have abortions rather than marry foreigners.'

The bishop puffed and protested.

'And you can keep away from the ceremony. I don't want you there. I only want your licence.'

It was a white wedding without a bishop. Without parents. Ariadne was a radiant, radical bride in a traditional white gown. The couple quit Boeing and settled in paradise. After Apostoli's birth Ariadne found a civilian job with the air force, employed as the first female engineer at the Souda Bay base.

'I was only a small mechanic,' she said, 'about four and a half feet tall.' With a baby.

In the midst of rebuilding an axial compressor she would put down her tools to look after Apostoli. Greek women had always fed their children in public and, as the government provided no child care, Ariadne felt entitled to bring her baby to the hangar, letting him play among the fan blades and breast-feeding him in a quiet cockpit.

'I fought to do what was only right,' she said, lighting another cigarette. 'Acceptance for what you are is like hot, fresh bread. You cannot eat it but it feeds you.'

Ariadne's husband was less of a fighter. And less fortunate. Being a foreign national and speaking dreadful Greek, he was locked out of Crete's two small aerospace companies. He made a start at various jobs including an air sightseeing company. But it relied on the Hania Flying Club's treacherous Cessna and, when it failed, he bailed out of paradise and into the arms of another woman.

'In those years my parents never once called to ask, "How are you managing? Do you need anything?" I didn't want sympathy. I always hated my mother's sweet-little-woman act, always wheedling, like a cat, so incapable. I hated all her theatre. I wanted equality in my relations. So I pretended to be like Tarzan, the strong one, until finally everyone believed it was true.'

Then, a few months before our arrival in Crete, her parents died. Her mother was diagnosed with cancer and wasted away in a fortnight. Her father, a hard man who had worked like a dog and never displayed affection towards his daughter, died seven weeks later of a broken heart. The loss devastated her, despite the years lived apart. Ariadne, who had been the hub around which lives revolved, making meals, debating, smoking, swearing, was alone.

'All I had left was my beautiful, stubborn boy,' she said. 'I went into a deep depression.'

Above our heads wheeling bats hunted for insects in the evening sky. Dogs barked at the darkness. Yióryio dropped a tray of glasses on the concrete floor. We finished our coffee in silence.

'In Crete we wash our dead with wine,' she whispered, bending forward, her voice full of emotion but controlled, 'just like it was before.'

Like the Minoans. And the Mycenaeans. And the Romans.

'I did this for both my parents.'

She pulled on her cigarette, nicotine-stained fingers turning beneath the pale lights.

'In her life my mother went abroad only once,' she went on, 'to Jerusalem to buy the clothes to wear in the grave. I dressed her in those touristy things, knowing that I would never touch her again.'

Greek women wear white only three times in their lives: at their baptism, their wedding and their death.

'We lay out our dead in our homes. In the evening people come and we talk about the dead and try to make them alive. I bought five kilos of good fish to make soup. You are forbidden to eat meat because the dead are of the flesh. Then we look at our dead, we cry, we go out, we eat, we drink wine and we don't sleep.'

Unknown birds shrieked in the night, shrews or rats rustled through the trees, raiding nests behind the *kafeneion*.

'My mother had skin like a baby. My father, whom I had always feared, was like an angel. All my life I'd known him only as a busy man who worried about our future. I never felt his soul until that night – but then he was dead.'

In the dark Ariadne quietly cried.

'That night my uncle he came late. He was a good man with a weak heart. He was weeping on top of my father and his arm went up and he knocked open an eye. Baff! Suddenly my father was no longer an angel but a dead, grey Cyclops. My uncle he called out to the Virgin. I thought he would have an attack, leaving me with another body in my hands. So I closed the eye of my father and the angel's expression was lost. But for one little moment I'd seen him as I wanted to see him.'

She toyed with her cup, her violet-green eyes downcast, alone.

'It is good sometimes to think that you have an angel next to you,' she said, 'even if it is an illusion.'

I wanted to believe her. I wanted my memories to be an enduring solace. I wanted to wake up in the morning and laugh.

'This flying is an offering to your mother, a driving into the bone,' she said. 'In Greece if you are in big need, then everybody is next to you. People they ring, they come and sit with you, they even sleep with you. That's how it is for us with Mr Death.'

And so it was that Ariadne came to me.

The Pano Anissari clock struck the hour, three minutes late. Behind us Polystelios dealt another hand of cards. Ariadne took my cup, turned it over and offered to read my fortune in the coffee grains.

'I see a journey with two obstacles, one at the very last minute. But you will overcome them and live. I think.'

I craned my neck to see the pattern. It looked to me like sludge.

'You see these two birds? What do you call them? Doves? One large and one small, coming out of an opening, out of a dark cave. The larger one is touching the smaller one with its wings, maybe even carrying it, like Daedalus and Icarus. Do you see them, coming into a lighter place?'

I couldn't see.

'This is a happy thing. It's very clear to me,' she said, putting down the cup. 'But of course the coffee is not dry so this is only a . . .'

I stared at her. I had been crippled by my mother's death. My salvation was a vision of white wings rising into the clear blue light. The flight I knew would be a ritual parting, a meditative rite. I wasn't off to fly alongside my mother's spirit, or to pluck one of Zeus's wing feathers. Rather, I was flying to take my leave of her, to free her by freeing my heart. To mark the moment and 'drive it into the bone'. Then I would come down and incorporate myself back into the earth, unless I killed myself in the process. It was an act of love that would be my grounding.

'It is time for bed,' said Ariadne.

'My poor mother,' I said to her. 'My poor, poor mother.'

Apostoli didn't arrive early on Saturday. In fact he didn't appear all day. On Sunday morning Katrin stayed in the house. I walked alone to the garage. Head down. Seeing only the dust.

'What you doing, *pilóte*?' yelled Yióryio. I'd thought the *kafeneion* was closed. 'It's Sunday. You must be in bed, or drinking wine, or both.'

'Today I will work,' I told him.

I didn't want to stay indoors, picking at food and falling asleep over a book. I had to fill the hollow days. But Yióryio looked so offended that I compromised.

'I'll just work for a little while,' I said, 'if that's all right?'

An hour later the *kafeneion* was packed with Sunday-best villagers. Relatives from Rethimnon and Hania were visiting for the afternoon. Children chased each other through the open doors.

At the bar I tried to order a coffee but no one paid me any attention. Yióryio was now too busy serving the noisy crowd. Sophia was in her kitchen preparing *mezés*. I turned to leave when Polystelios hailed me to his table.

'Sit,' he instructed.

I managed to catch Yióryio's arm and ask for coffee.

'No coffee. You drink wine.'

Polystelios was dressed in grey and black, as were Socrates and Kóstas. He had even cleaned his nails.

'There are three Anissaris,' he reminded me, his voice rising an octave. 'Pano is good and Kato is hard-working but Messa has the best wine.'

Food began to appear as if by magic. I'd heard no one order it. Polystelios and Papoos dished out servings of whitebait, *spanakópita* and platefuls of *hórta*. Children distributed saucers around the other tables, including to their mothers who sat together in the corner with Aphrodite. I tried to explain that Katrin was making us a soup at home so Leftéri was dispatched – at the controls of his new F-16 model – to fetch her. When she arrived, Yióryio put a two-kilo grouper onto the barbecue. Sophia produced fried potatoes swimming in olive oil, and lettuces, picked minutes before in their garden.

'*Kalós orísate*,' Manólis recited in welcome.

One of the village boys sat beside me, pushing aside the plates, wanting help with his English lesson. He was captain of the senior volleyball team. I asked him his plans after finishing school. 'Will you leave Anissari?'

'Leave?' he repeated, as if not understanding the word. 'Why? I will stay forever.'

'This my philosophy too,' said Yióryio, putting his hand around me. 'It not possible to go all places so I stay here.'

'Anissari is not too small?'

'Anissari is too big,' he laughed. 'It is big village and you will buy a house here,' he insisted, his face set askew in its pleasing squint. 'You start with one room, one olive tree and a goat . . .'

The starters were followed by salvers of *pastítsio*, *yemistá*, *yíyandes*, more than anyone could eat. I tried to stem the flow of food to our side of the table but was told that we were guests and must eat. *Filoxenía*. There was no question of refusal. Or payment. It wasn't a saint's day or anyone's name day. It was just a Sunday.

Until that day I'd never considered that the villagers might be responding to my need, that their care and enthusiasm was relative to the depth of my mourning. As Yióryio often assured me, Greeks knew everything, which was far more than I did. I didn't even know how soon death would come to Anissari, and how my white winged story would soften another's pain.

'*Kalós sas vríkame*,' I told my hosts. It is good that I have found you.

I did not finish my hardware shopping list. Or write up my notes. The afternoon vanished in filling our bellies. And softening our heads. It was delightful.

'In your book write please that you have good neighbours here in Anissari,' asked Yióryio sometime after four, reaching again for the hospitable flask of *tsikoudiá*.

And I did. It was true.

11. Trouble in Paradise

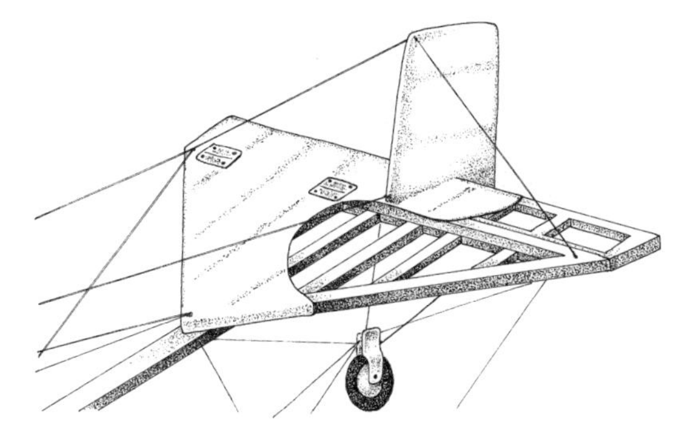

At dawn on Monday Polystelios cut bamboo to build frames for his runner beans. Ulysses craned open his lazy eyes and strode off to destinations known. The village children huddled outside the yawning *kafeneion*, a clump of sleepy heads and artificial fibres waiting for the school bus. While in the garage Katrin and I began the ironing.

The Woodhopper was in a state of undress. Its wings were wrapped in Ceconite but the material hung on the frame like a décolletage frock. If it was to fly, my aircraft needed to wear a skin-tight leotard.

Ceconite was a heat-sensitive fabric. At 250°F it shrank by 5 per cent. At 350°F it contracted by 10 per cent. At 450°F it burst into flames. A controlled heat needed to be applied, which scuppered the idea of using Katrin's hairdryer. Its temperature was erratic and I didn't want to run the risk of overtautening – or melting a hole in – the material.

The haphazard instructions suggested using an electric iron, its temperature calibrated with a sugar thermometer. No one in town owned a thermometer apart from the vet, and his was best suited to a sheep's backside. But I had brought from England my mother's old iron which, having pressed my school shorts in days past, would now tauten my wings.

I set the temperature gauge at the 'wool' setting and hoped for the best. I glided it over the left wing, without pressure and with the face plate only just in contact with the fabric. At first nothing happened. Then the Ceconite crackled and sprang upwards like a hot air balloon. In less than a minute the bay closest to the wing root had shrunk to a drum-like tautness. A drachma coin bounced off its surface.

Next we ironed the wing tip and then the second panel in from the root, working back and forth across the upper surface towards the centre, balancing the shrinking load. The wing ticked and creaked with the increased tension and the sound delighted me. Katrin smiled sadly.

Apostoli still hadn't arrived by ten o'clock, or eleven, so we worked on, turning over the wing to iron the bottom bays. The flat underside creaked less than had the curved, upper surface.

At one o'clock, as we stopped for lunch, Apostoli soared into the village square. Sophia's chickens fanned out of the path of his silver Fiat. The Olympian strode up to the garage door wearing a golden flight suit, raised his shades onto his gelled hair and said, 'Good morning.' He tipped his shoulders in a mock bow and shook my hand. 'I have arrived.'

He might be two days late but at least he had come.

'How was your evening?' I asked. Or evenings.

'Amazing,' he said, rolling his eyes. 'It is the start of the hunting season.'

He didn't mean for small birds.

'Women come here from all of Greece as Cretan men have

a reputation.' The Greek flag was stitched onto his shoulder. 'It may not be true but who am I to disagree?'

I ushered him into the garage. He seemed surprised by the speed of our progress. The ironed left wing appeared to be sturdy and solid.

'Doesn't it look beautiful?' I said.

'Yes,' replied Apostoli, narrowing his all-seeing eyes, 'but it is not correct.'

'Not correct?'

'It is bent.'

I knelt on the floor, stared along the leading edge and saw the pronounced upward bow. As Katrin and I had ironed the upper surface the fabric had pulled and warped the wing. It had not occurred to us that the cracking sounds were of the frame being distorted.

'You have changed the aerodynamics,' said Apostoli. 'Maybe it will not fly.'

I didn't have enough Ceconite to re-cover the wing. In any case the ribs and struts may have been damaged.

'But it is no problem,' the Oracle announced. 'We can iron the lower surface again and that will straighten out the bow.'

We flipped the wing, reheated the iron and tried to tauten the underside fabric. It didn't work.

'The problem is that you are building an aeroplane,' said Apostoli, 'and that is difficult.'

The correct way to iron a wing – Apostoli instructed us over his second iced frappé – was to do one panel at a time, top then bottom, flipping the wing again and again. Then the tension would be balanced and one surface would not pull against the other.

'It is not a bad wing,' he said. 'It is a wing.'

We discussed whether to iron the second wing 'incorrectly', as we had done with the first, and have two bowed but balanced wings, or to iron it Apostoli's way and have an unmatched pair.

'You have forgotten about the wires.'

'You said that there are too many of them.'

'Not now. Now they will work for us.'

The finished aircraft was to be held in tension by dozens of wires running between the wings, fuselage and tail. So although we were unable to fix the first wing, we convinced ourselves that its bow would be corrected in flight by the wires. Or by gravity. Or by luck.

We worked through the afternoon without speaking, turning over the wing after ironing each panel. Apostoli was careful and conscientious.

'Double your eyes,' he said to me, again and again, meaning be attentive. *Ta matya sou tésera.* 'We will do this without a thousand difficulties.'

'Why are you so anxious today?' he asked me around six o'clock.

When we finished, the second wing seemed to be balanced in exquisite tension. We laid it beside its pair and bent down to compare them. We doubled our eyes. Both wings had an identical bow.

'I don't understand,' I said.

Neither perhaps did Apostoli. But he said, 'Now we have no problem.'

Apostoli didn't turn up the next morning. To him time-keeping was a concept, more a matter of interpretation than doctrine, and he had little inclination to submit to its rule.

Katrin and I worked on alone, assembling the rudder and horizontal elevator. The tail was to be made of ¾-inch timber, which seemed flimsy in comparison to the wings' long, heavy struts. Its slender secondary braces, made of light polystyrene to reduce weight, suggested to me that the whole structure might disintegrate in flight, ripping off the Ceconite in the process. I remembered an anxious dream about my maiden flight. In it I climbed into the clear blue sky as the wings begin to tear. My aircraft was stripped down to its bony frame and left suspended in space like a tiny cartoon. Tatters of fabric fluttered in the air currents behind me. I fell into the sea with a plop.

So I tried to imagine where the structure might break, inducing a spin or stall, and strengthened the stress points. I laid strips of fibreglass onto hinge and bracket points, and places where the anchors – with their structural wires – would be attached.

By early evening we completed the reinforced frames of both the rudder and elevator. It was a satisfying day until I discovered an oversight. In the plans insufficient clearance had been allowed for the rudder. If built as designed the tail would jam against the end of the boom, preventing me from turning left or right and sending me straight into a mountain. It needed to be re-devised.

The next morning Apostoli arrived on time.

'As you can see I am here,' he announced with a customary bow.

Together we began to cover the tail, cutting the fabric for the rudder and imagining the flow of air across the elevator. I wore polythene gloves to mix the adhesive and acetone. The smell of chemicals mingled with his aftershave. We brushed the struts with the cement then laid the cut fabric onto the frame, working in the glue with brushes and our fingers. The acetone dissolved the tips of the gloves leaving telltale plastic fingerprints all over the tail.

While the adhesive set Apostoli took a cigarette from his

pouch and smoked in the square. Nicotine seemed to be his main source of sustenance. He never appeared to eat and he drank only the occasional Coke or coffee. We invited him to the house for lunch but he preferred to stay at the garage.

'Leave me here in paradise,' he said, taciturn, all-knowing, sitting cross-legged on a table watching the glue dry.

In the afternoon a jet fighter shot down the length of the valley, a silver silhouette above the donkey trails.

'A-7H,' Apostoli said, sparked out of his reserve by a flash of excitement. 'The flying is magic.'

'You've flown a Corsair?' I asked. The 340th Squadron was based at Hania.

'Only on simulator. My mother arranged it last month. I did two loops over the runway. I want to visit a lot of places,' he went on, 'but when you are flying you are flying. So the place doesn't really matter. Only being topside is important.' He stubbed out his cigarette and doused the butt. 'In the end I will come back to Crete because I love my place. I love the rocky mountains, the people and all the history in this island.'

'And the crazy nights?'

'Especially the crazy nights.'

While the light lasted Katrin and I trimmed the frayed fabric edges and loose strands. Apostoli covered the subfin, working the fabric around its butted edges, inch by careful inch.

'You could make a pleat here,' I suggested to him.

'I have a plan,' he assured me, which meant he hadn't made a decision. I was learning that he tackled problems only as they arose. 'I have a better tool for this at home' translated as 'I can't deal with this until tomorrow.'

Apostoli completed the fin in silence, waving the scissors in his gloved hand, ladling glue onto the fabric. He was excited by his work, even though he botched a corner and I had to remedy it later that evening.

'Be happy,' he told me. 'I like this tail very much.'

Over the next two days we painted the wings and tail. The conventional method for sealing aircraft fabric called for three cross coats of nitrate dope followed by three fill coats of clear butyrate. As my Woodhopper was to fly only once I used a single tin of latex emulsion, smoothed on with rollers.

The interior white house paint created an illusion of completeness. Villagers stopping by the garage once again expected the Woodhopper to fly by the weekend, even though we had yet to build the fuselage. And had no engine or runway.

'Two tickets for Heraklion please,' said Polystelios.

'How much is it?' asked Papoos.

'One way or return?'

'One way.'

'Very expensive,' I told them. 'Ten drachmas.' About a penny. Papoos offered me a coin. 'For pensioners it's free.'

'Then you can make it a return.'

The first tourist of the season alighted in Anissari like a migrating Aryan bird, wearing sandals and white socks. '*Was machen Sie hier?*' he asked. 'You are making a *Flugzeug*? An aircraft?'

I explained.

'Why not? Some people collect stamps.'

Other holiday-makers followed, snapping photographs in silence before driving onto the beach. One alpine-cold Austrian couple had no doubt that my objective was suicide.

'I hope you have lots of life insurance on him,' the husband told Katrin. 'If you take out many policies you will never have to work again.'

Apostoli came to the garage most days, attaching bolts and anchors and persuading me to buy a mobile phone 'so we can work with more efficiency'. I began to tolerate his contradictions, a meticulous professionalism without responsibility, an assiduity combined with an inability to get out of bed. He claimed to have found a spare water pump motor which would suit the Woodhopper, given that the prototype had been propelled by a chain saw. He also promised to borrow a hot knife to slit guy wire holes in the fabric. But the next morning the Oracle sent me his first text message. 'I cannot come today. Have stomack problem. I send my mother for the knife.'

I borrowed Yióryio's sheep-branding iron instead.

That Thursday he again arrived late and filled with good humour. He had met a Swedish tourist and stayed up all night with her discussing the meaning of life and location of the best beaches. I had been counting on his help to lay out the aluminium cockpit.

'You must get an understanding of Greek time,' he told me as I worried about the schedule.

'I'm just trying to get the job done before the winds.'

'If you don't fly this year then there is always the next,' he shrugged. 'If you want punctuality go to Zurich.'

Then, on Friday, the demigod arrived both late and in a bad mood.

'I have a problem,' he growled after finishing his frappé.

No engine? No airfield? No Swede?

'Tonight is Friday and as you know Fridays are crazy nights. But tonight I want my girlfriend to go home.'

'You want to take her home with you?' I asked. I didn't doubt that he could proffer an attractive proposal.

'No, no. I want her to go to her home alone.'

'While you stay out?'

'Of course.'

It transpired that he had promised to go dancing with the Swede, as well as with his Cretan girlfriend.

'But what if your girlfriend decides to stay out too and finds you?' I asked.

'She will say, "You here? I kill you." She is a Greek after all so she will make trouble.'

'Not only Greek women,' said Katrin.

He sighed an Olympian sigh. Apollo had also had trouble with his love affairs.

'I must think of some story at the moment. Or not at all.' Apostoli toyed with his mobile. 'If you will excuse me I will sit in my car.'

He spent his afternoon there, making calls, adjusting his arrangements, spinning stories – and agreeing to meet us in Hania on Monday with his motor and mother.

Until he changed his mind.

'Tuesday is better,' he said, now lighter of heart.

'Why?'

'Because Monday morning comes very soon after Sunday night,' he said with a twinkle in his black eyes. 'A Greek for Friday. A Swede for Saturday. As for Sunday, the gods have it. Amazing.'

12. Little Hairs

I laid out the aluminium with the help of little Leftéri. He put aside his latest flying machine, a Heath Robinson affair assembled from offcuts and spent shotgun shells, to hold the axle while I positioned the cockpit pieces. As I worked he whistled and chatted with his imaginary birds.

'Come for coffee at my house.'

I looked up. At the garage door a raven-robed figure flashed an easy false smile.

'I'm busy at the moment, Papá Nikos.' We'd had few dealings with the priest.

'Come now and I will tell you a story which you will never forget,' he insisted and walked away.

I sighed. I laid down a tube. I counted to ten. Leftéri picked up his flying machine, wiggled his fingers under his chin and ran home laughing.

★

I distrusted the priest, though not merely because his shifting eyes looked like black searchlights, scanning the skies for low-flying opportunities. Rather, my unease arose from his gold Rolex, flashing its twelve-carat face from under a frayed sleeve, and the new smoky Opel in which he lorded it from church to church. Both luxuries seemed incongruous with a life of humility and self-denial.

Nikos lived at the edge of the villages, within them yet set apart, in a plain, ageless room with uncounted Lego-block concrete additions. Around the house jacaranda trees and yellow mimosas thrust themselves into bloom. Inside there were three tables under a faded Olympic Airways poster of short-skirted flight attendants. The tables were laid with airline cutlery. In the corner was a cooker. A mobile phone leaned against an icon.

'Greetings, friends,' said the priest as Katrin and I arrived, lifting his arms above his head in exaggerated welcome. He gestured for his dormouse wife to serve us and ordered a young woman, who was heaping garden clippings over a wall, to carry a table out under the vines. His son sat on the sofa watching television and yawned.

'Please sit down,' he said. His pencil-thin eyebrows met in a knot beneath a skullcap of colourless hair. 'In summer we serve lamb on the spit but now we will drink coffee together. You are my guests.'

In season the room, their living room, doubled as a restaurant for passing tourists who were caught in his searchlight. Nikos drank with them, told them his story and ordered his wife to bring them more chips, more bread and complimentary flasks of home-made *tsikoudiá* until, unsteady on their bicycles or unable to drive back to their hotel, he put them up in his maze of upstairs rooms. For a modest fee.

'All guests like to eat and stay very much,' he assured us, leaning forward to add, 'they find it excellent value for money with breakfast included in the price.'

To prove his point Nikos produced two large buff envelopes stuffed with letters and cards from Bremen, Bordeaux and Basildon.

'These are from the last six months only,' he crowed, laying before us enthusiastic recommendations from holidaying Germans and Scots. We glanced through the correspondence. Monique raved about his olive oil. His Greek salad was 'the tastiest on Crete'. Another tourist wrote, 'You make us feel so welcome with your hospitality and your fantastic story.'

Yet still I didn't trust him.

'Are you married?' he asked us. 'If not then I will marry you this summer, without charge because we are friends. I have four churches. Choose the one you like then after the ceremony your family and friends will eat here in my house.'

At a profitable 5,000 drachmas a head.

His wife brought us sweet, grainy coffee in a tiny Olympic cup and saucer then withdrew to join her son on the sofa. He took her hand. Her drab, brown mouse clothes looked decades old. Nikos' patches showed under his cassock, as did his gaping button fly. Each summer he put on weight but refused to spend money replacing trousers that would fit him again in winter.

'You will eat with us,' the priest told us. 'Every day you are welcome.'

We thanked him for his invitation.

'My son will join us then, if he has time.'

By way of an introduction the son announced in English, 'I love my father and my mother but my wife is . . . my wife.'

The young woman, the boy's wife and Nikos' daughter-in-law, now stood at the sink peeling potatoes.

Nikos was the priest for Pefki and Paleloni, hamlets in the hills behind the house, as well as for the Anissaris. We had seen his stark white chapels perched on green hilltops ringed by Cretan cypresses. I remarked on their beautiful location.

'They are too far apart,' he replied, wishing for an easier life. 'Unless you can fly me between them.'

I asked if any of his churches was dedicated to Agios Elias. The prophet was said to be the Christian successor to the sun god Helios. When Christianity rolled across Greece Elias had inherited his mountain sanctuaries and sun-towing, four-horse chariot.

'We don't speak today of the idolatry of the twelve gods,' answered the priest, appearing to dismiss the survival of ancient rites. 'But to fly means something special to us.' He snapped his fingers and the young woman set tumblers of water before us. 'It does not mean that a man is clever, only that he has faith. It means he has the ability because he *believes* that he can fly. You know, of course, the recent story of Piathus?'

I didn't know it.

'Piathus was a flying monk who soared from mountain to mountain tending to the churches around Mount Athos. I consider myself to be like him,' he confessed, 'even though I need metal wings. Piathus and I share faith and a belief in miracles. The neighbours will deny this because they are jealous.'

I asked him why he had become a priest and his soft, loose cheeks rose with his smile. 'It is a fantastic story of two miracles,' he assured me, casting a glance towards his wife. 'You will write about me in your book so everyone will know the truth.'

'The truth?'

'That I was the first man in Anissari to fly.'

At the end of the 1950s Nikos had been a young conscript, working as a deck-hand aboard a naval supply vessel.

'A strategic ship,' he explained as his daughter-in-law brought a neat plate of tomatoes and soft feta. 'We delivered torpedoes for our navy to fight the Turks.'

On leave in Athens one summer weekend he was bored, killing flies, when a shipmate told him about the 'Airline Days'. To encourage air travel the newly formed Olympic Airways offered Greeks the chance to have their first taste of flying – for free. Even as a young man Nikos could not refuse a bargain. He hitched a ride to Hellenikon airport and boarded a silver and blue DC-3 Dakota, the airline's single aircraft.

Olympic, founded by Aristotle Onassis in 1957, was not the first Greek airline. Icarus, the Hellenic Airlines Company, had become the country's earliest civil carrier in 1930. But its primitive equipment and rudimentary airports ensured that the public kept its distance. In under a year Icarus came down to earth like its namesake. Various other operators followed including Technical Airline Exploitations and, after the war, ELL.A.S., a collaboration between the Greek armed forces, the civil service and Scottish Aviation. ELL.A.S. pressed second-hand Consolidated Liberator bombers into passenger service. Another company named Daedalus flew a decrepit Junkers 52. Its licence was revoked due to technical failings, or perhaps national pride – only a few years earlier the same aircraft had dropped Nazi paratroopers onto Crete.

'The airline was called Olympic after the celestial home of the gods,' said Nikos, then added as an afterthought, 'in the dark years before Christ.'

In the Dakota he sat between a stern priest and a young Athenian in a printed frock. The earth fell away and the aircraft climbed out over Glyfada, banked above the Savonic Gulf then circled the capital. From 2,000 feet he saw Piraeus, first fortified by Themistocles in 480 BC, and his own freighter at anchor in the Great Harbour. He leaned across the woman, excusing himself, to see Aegina, blue mount on a silver sea. She smiled at him as they turned back towards the city. There was the Hill of the Nymphs and the Agora, Syntagma Square and the bald cap of the Acropolis. His eyes darted from the porthole window to the woman's knee, left uncovered in her own excitement. The Parthenon slipped across his vision as the sun caught the soft hairs on her legs.

In that moment the Dakota hit an air pocket, as aircraft do, and fell 500 feet. The passengers, who had been enjoying the excursion, began to scream. The priest called out to the All Holy Virgin. The woman grabbed Nikos' hand.

'I'm frightened,' she yelled to Nikos.

'We're frightened, Papá,' Nikos shouted to the priest.

'Pray, my son,' he advised.

But the Dakota didn't stall. It didn't crash in a fireball into the Temple of Olympian Zeus and consume three dozen lives. Rather, it shuddered back into flight, as aircraft do, dropping its nose towards the runway and returning to the earth with an easy three-point landing. But in the incident Nikos saw a miracle, especially as the young Athenian hadn't let go of his hand until they reached the tin terminal building.

'My destiny was set in that moment,' he told us.

The three elements that would dominate the rest of his life had come together in a ten-minute sightseeing hop.

'I understood the beauty of the world, of women and of prayer. It was a miracle; the first miracle. Have another drink.'

As soon as Nikos completed his military service he applied to join the airline. Because he spoke English he was hired as a cabin steward and assigned to the Athens–Rome–Paris–

London route, inaugurated with the delivery of a nearly new DC-6B.

'It was difficult work,' he told us, and not because of the noise or the hours. 'How can anyone concentrate on serving coffee while flying over the Alps?'

When not checking seat belts or handing out barley sugars Nikos was at a window, gazing down at the Vatican, Le Bourget or a pearl-grey Thames snaking through the sprawling green city. In London the crew bunked five to a room at the Bayswater Hotel, their devalued drachmas too paltry to buy more than a pint or pie, which didn't excite their palates in any case. So rather than go hungry, or stoop to in-flight *coq au vin*, the stewardesses took to stowing on board *dolmádes* and *keftédes*, cooked at home by anxious mothers and eaten off embroidered Cretan tablecloths in Hyde Park. Sweets were provided by the co-pilot's brother, a Monastiraki baker reputed to make the best *loukoumádes* in Attica. The pilot brought bottles of wine made from his brother's vines in the Peloponnese. Nikos began each meal – as he did each flight – with a prayer.

Greece was tumbling out of purdah and into bed with Europe. Freed for a day or two from the orthodoxy of home, upgraded to good hotels with Onassis' continued investment, the young men and women of Athens and – apparently – Anissari found themselves released, exhilarated, sharing two to a room. Overnight stops became salacious adventures. On his name day the co-pilot treated the crew to dinner at Ronnie Scott's. The chief steward danced the *zembékiko* under Marble Arch. The stewardesses' elaborate, uniform headscarfs, which harked back to the Ottoman veil, were cast aside on the thick pile of the Palm Court Hotel's carpet.

'I managed all the sins in London,' relished Nikos without

a hint of remorse. 'I felt pride in the blue and silver uniform. I lusted after women and revelled in them. And I avoided handing out hot towels as much as possible so I could sit and enjoy the view.'

He was the first priest I'd met who bragged about temporal exploits.

'We flew. We lived. And, praise God, there were no unwanted babies. Every morning I assembled the crew on the tarmac under the nose wheel to pray for the miracle to continue.'

Throughout this far-fetched story Nikos' family paid him little heed. They'd heard it all before. The son fell asleep on his mother's shoulder. The daughter-in-law finished peeling potatoes and began to wash the greens. The cat dropped a dead mouse on the tile floor.

Some years later Nikos met again the brown-kneed girl who had grabbed his hand on the Airline Day. The two young people recognized each other while gathering wild flowers along the runway to deck aircraft noses on May Day 1966. That first Dakota flight had also moved the woman to join the company, after she'd persuaded her mother that Olympic wasn't a winged sin bin. Which, of course, it was on international routes. To her mother's relief her first posting was on the domestic Thessaloniki service. Even in those early, starry-eyed years the route was something of a bus run, out and back twice each day except Sunday with no overnight stay. Every evening she returned home for supper. While Nikos looked after elite travellers, pouring Samos Muscat into crystal glasses and stowing Hermès hat boxes in the cockpit, the brown-kneed girl chased escaped chickens down the aisle and humped kegs of pesticide into overhead lockers. Her poorer passengers crossed themselves at every hint of turbulence and wondered if they'd ever see their fields again. She once persuaded a Sfakian shepherd to put away his revolver. He'd refused to submit – as had many other Cretans – to the indignity of a fastened seat belt. On another occasion a passenger from Thrace asked her to explain the function of the overhead call button.

'Use it to call for wine whenever you want,' she said. Later as the chime rang again and again, the passenger was found holding an empty glass under the call button.

'I pushed it,' he complained, 'but no wine comes out.'

Then in 1966 Onassis bought the first Big Jet, a Boeing 707, and extended Olympic's network beyond Europe with new routes to America and Australia. He introduced gold-plated cutlery in first class, commissioned Pierre Cardin to design crisp air crew uniforms and sought to install an in-flight piano for the entertainment of his full-fare business passengers. The Big Jets launched a new era of air transport, though not just for the wealthy. With twice the capacity of the piston-engine generation and flying twice as fast, the 707 reduced costs and brought air travel within the reach of the general public. Economy fares broadened travellers' horizons, reuniting families and friends at Hellenikon, Idlewild and Sydney's Kingsford Smith.

Both Nikos and the brown-kneed girl were assigned to the inaugural Australia flight. In the depths of the night high over the Arabian desert they made love in the aft toilet.

'My lucky thirteenth stewardess,' he boasted to us on his shady terrace.

'Only thirteen?' I asked.

'In the air.'

After their cramped copulation Nikos stared out of the window at the darkness below, watching for the twinkling spark of a Bedouin's fire, whispering a prayer as he held the girl's hand.

Olympic became 'a company of five continents', flying 707s on long haul, 720Bs on high-traffic domestic routes and twin-engine YS-11 turboprops between the islands. But the time had come for Nikos to set foot to ground. He had saved enough money to enlarge the old family house in Anissari. He married the brown-kneed girl, who sat across from us in cloth slippers watching television, and as every Cretan must do he returned to his village.

'I had seen the world. I had slept with beautiful women. And I returned to find only jealous neighbours. I tell you my friend, jealousy is the greatest force in Crete.' He leaned forward to hiss, 'In Theriso a shepherd was shot dead because of his success with women. Lucky for me the Anissariots are cowards and do not shoot. Instead they have poison tongues. In five years no one called on her,' he said, angry now, flicking his hand at his wife. 'Every evening she waited here alone. It cut her so hard that she could not give me children. Not even a girl. I, who had served a bread roll to the Patriarch, lost face.'

For five years the brown-kneed girl had remained childless. She'd suffered from abdominal pains since moving to Anissari – at first the cause was thought to be the change of water – and consulted three doctors plus an infertility specialist in Heraklion.

In rural Crete a couple without children was like a tree without fruit – worthless and wasteful. Nikos, as dry as a failed farmer, withered in the eyes of the villagers. His prayers went unanswered.

'I was not a superstitious man but I began to think of my past sins.'

Then one night in summer Nikos had a dream.

'Never in my life did I remember my dreams,' he said. 'But this once I woke crying like a baby. I was trapped in a burning plane, falling onto Anissari.'

In his dream the aircraft – which he said was a Comet – crashed behind their house, burying itself in the loose, red soil of his vineyard. In the dark Nikos shook awake the brown-kneed girl, made her light the lamp and follow him with a shovel. He dug at the spot shown to him in the dream. He fell on to his knees and clawed at the earth like the village dogs which now howled at the night. His fingers found an object, familiar to the touch yet out of place. He dragged a crumpled plastic box from the hole. He brushed away the dirt.

It was an Olympic Airways economy class meal tray.

'O seed of Charon.'

Nikos flipped off the transparent cover. He opened the tin-foil main meal dish.

'There it is,' whispered the brown-kneed girl.

Rather than rotted chunks of *souvláki* or a tossed salad with shrivelled prawns, inside the dish was her bridal shoe pierced by a nail. Beneath it lay a dried end of soap stuck full of pins, tangled combings of hair and a padlock.

Nikos' bride had been 'nailed' by a pernicious villager.

He and the brown-kneed girl carried the tray into the house. For all their jet-setting modernism they knew how to deal with archaic, common curses. By the light of the hissing lamp they pulled the nail out of the white shoe. They unpicked the pins from the bar of bridal-bath soap. They straightened the knotted hair, which may have been her bridal combings. Then at dawn the priest took the first bus to Hania, found a locksmith to open the padlock and threw it into the sea.

In an hour the woman's pain left her.

'That day I resolved to take their stupid old ways from them,' seethed Nikos, who had no doubt as to the identity of the villager who had cast the curse. 'I would save them from their pagan folklore.'

Within the month he had joined the priesthood. He took

his vows and with the help of the Bishop of Rethimnon, who had once been a frequent traveller, he was given the parish of Anissari.

'I had to show them that, in the eyes of God, they were not better than me.'

His eyes blazed with an inner flame, though it may have been heartburn.

'This was the second miracle?' asked Katrin, as incredulous as me.

'No,' he answered. 'The last and best miracle was the birth of my son. Nine months after my dream.'

Nikos touched my shoulder, refilled our tumblers and said, 'That is the story of my remarkable life. 'So when your book is published you will pay me for it.' Then the fervent, fat little man left us alone.

Across the lane his daughter-in-law moved a goat out of the sun, untying it from a stake in the long grass and tethering it to the trunk of an olive tree. The animal craned its neck to nibble the swaying branches. Pick-ups returned from the fields heavy with oranges. A vegetable vendor crawled through the village, puffing though his tinny tannoy 'Artichokes! Fresh artichokes!', stopping to sell the long-stemmed crowns from his tail gate. The bad-bread van tooted its horn and the priest bought three loaves. The son sat on the sofa eating a plate of chips fried just the way he liked them and watched television.

In the evening we carried Aphrodite into the fields to pick camomile. As I panted over the rutted earth butterflies rose up before us. She cooed in my ear that all her lovers had been stronger than me. And better looking. She weighed a ton and I was tempted to drop her into a cistern. Instead I sat her on a wall from where she directed our efforts and, after I'd caught my breath, we stuffed the golden flower heads into supermarket carrier bags.

'You spent the morning with the priest,' she said. She knew the news from Sophia who had been told by the bad-bread baker.

'He invited us for coffee.'

'He is a thief,' she spat with unexpected anger. 'Once he asked three French tourists for a meal. They said to him, "Is this an invitation?" and he told them that they were his guests. So they ate and when he asked them for money they refused to pay. Papás told us later to watch out for them but we all laughed at him. How much did he charge you for lunch?' When I told her she said, 'Be careful of him. He is always after something else.'

'He told us his story,' said Katrin, 'which was remarkable.'

'What's remarkable about him? He grew up in that house. His family had nothing. He married a nobody. Now he steals from everyone.'

'He told us how he became a priest,' I persisted. 'About flying and about the miracle of their son's birth.'

'He became a priest because he didn't have the brains to do anything else. And the only miracle in his life is that he managed to impregnate his rat of a wife.' She lowered her voice. 'The doctor told me that he is impotent.'

'The doctor told you?'

'Or the nurse. Do you know why their boy is like Christ?' she asked us. 'Because he lives at home until he is thirty, he believes his mother is a virgin and his mother thinks that he is a god.'

'So Nikos never visited London?' asked Katrin, a hint of disappointment in her voice.

'He has slept every night of his life in that chicken shed,

apart from the day he went to Hania to buy his car and argued until dawn about the discount.'

To call someone a liar Greeks wiggle their fingers under the chin. The gesture is called 'little hairs' because priests have beards and are always believed to stretch the truth.

Maybe.

Or maybe not.

Evening had crept upon us, casting the world into lavender shadow. The sun dipped below the far peaks and its last beams caught the tops of the olives. The voices of the village children rolled over the banks with the cool night air and Polystelios called his sheep down from the hills. He milked them in the dusk under the trees while the island slipped into darkness, as if swallowed by a whale.

'My tears have watered these hills,' moaned Aphrodite. The rawness of her features had softened in the twilight and she sighed like a young calf. 'When will you take me to fly, *pilóte*?' she asked me. 'I can't wait forever.'

13. Loop the Loop

Every family in Crete owned a terrace of olive trees, a vineyard or a whitewashed farmhouse in the mountains. Every Sunday most city-dwellers returned to the land.

'To do what?' I asked Apostoli.

'Nothing special. Just to visit our place. Just to be Cretan.'

Lawyers turned the soil. Web designers collected snails. Academics spread dung. Even the most urban family was not long off the land.

I needed to find hardware. Plus records of early aviation. Katrin and I drove forty minutes west to Hania, a single bolt clutched in my hand. We parked next to Minoan excavations, breached the medieval walls and walked by a Turkish prison to reach the Venetian harbour.

Hania revealed itself as a capital of light and shade. Sunshine flashed off its half-moon harbour, flinging mackerel patterns onto its sea walls, dazzling the drinkers in the cafés which ringed the port. Ancient fishermen cast their lines into the glinting waters, tied a nylon end to quayside benches then retreated under pastel awnings for coffee. Above their heads ochre shutters swung shut against the late spring sun.

The searing light barely penetrated the narrow back streets, laid out by the Venetians and unchanged for five hundred years, despite the threat by a recent mayor to bulldoze the buildings into the harbour and create a piazza with his statue at its centre. The cobblestone lanes sloped up from the water, curving with the contours of the headland, overhung by rusted balconies and lines of clean washing. Crumbling houses rose three or four storeys from vaulted ground floors to lofty double-height ceilings. Vines had taken root in their aged buttresses and fig trees grew through broken fanlights. In parts of the old town the stonework seemed to be held together only by butter-thick lime wash. Some walls were lonely façades, without exterior rendering, their balconies propped up by timber jacks, their upper floors collapsed and stinking of pigeons. Palm-sized cakes of paint snapped away in our hands, each coat of white, terracotta and Venetian red taking us back a generation.

The city had been bombed in the Second World War and half of its medieval houses destroyed. In 1941 German Junkers 88s had flown in from the west, over the stubs of stone from which a protective chain had been strung across the harbour mouth since Classical times, blitzing as many as thirty Venetian *palazzi*. A decade later an English painter had found amidst their rubble shards of plump Minoan pottery, the first teasing remains of an earlier devastation. A clay tablet fragment was unearthed which read 'ten pairs of wheels (for chariots)' in Europe's first decipherable language. Another Bronze Age tablet, baked and preserved by a previous catastrophic fire, said, 'This town was destroyed by a violent conflagration.'

The island's history was a story of occupation and defiance, destruction and resurrection. Crete's cities were razed only to rise again.

Hania had been in continuous human occupation for six thousand years, longer than any other city in the Western world. Neolithic man carved caves in the limestone at the foot of Kastelli Hill. Above them on the acropolis Minos' grandson built Kydonia, naming his palace and province after the quince, Aphrodite's sacred symbol of love. Hania's apples were sold in the markets of Jerusalem at the time of Christ. The ship carrying St Paul to Rome wintered on the island about AD 60. The Byzantines, who inherited Crete from Rome, dismantled the city's ancient temples to build its churches and defensive walls.

In the ninth century Andalusian Arabs had given the harbour its modern name, el Han, meaning refuge or stopping place, although modern-day nationalists insisted that Hania was derived from the Greek word for vegetable patch. In the second half of the thirteenth century it was renamed La Canea when one in six Venetians immigrated here to construct 'Venice of the East', importing cosmopolitan sensibilities, stone corbels and Serbian mercenaries to man the garrisons. Hania became a centre of authority and wealth whilst the surrounding countryside was cast into poverty and revolt. In the sixteenth century the island's population all but halved through emigration and rebellion.

'Better the turban of the sultan than the tiara of the pope,' cried the Cretans when the Turks laid siege to the harbour in 1645. Although the Ottomans allowed Greek to be spoken and the Orthodox religion to be practised, they imposed their authority with a scimitar brutality which in time turned the Cretans against them too. The sultans also proved to be miserly administrators. In their two centuries of stewardship they built only a handful of mosques and five miles of paved road. They left Crete in 1898 when the island was placed under the governorship of the Great Powers.

Fifteen years later the Greek flag was raised over Hania, marking the end of two thousand years of occupation. With the deportation of thirty thousand Cretan Muslims in the early 1920s and the murder of the Cretan Jews by the Nazis in the Second World War, farmers and shepherds came out of the hills and settled in the vacated city. No pure-blood Venetians or hand-me-down Turks remained, the hunger for revenge having been too violent for genetic continuity.

Katrin and I stepped between their whitewashed squares and dark alleys, from the Levant to Europe, between the fifteenth and twenty-first centuries. The bombed Turkish steam baths had been reborn as guest houses and a restaurant opened in the shell of a burnt-out soap factory. In the walls of Venetian *archontika* souvenir shops sold Icarus mobiles, Italianate door knockers and *Wanderungen auf Kreta* guidebooks. An elderly matron opened her parasol.

Everything flourishes on Cretan soil: avocados, families and concrete houses. Between a rank of botched apartment buildings and lush allotments stretched the Minos Street farmers' market. In the open air stallholders sold sweet persimmons, warm eggs and live chickens. There were thumb-size local bananas which tasted of lemon, and fresh octopuses laid out on ice, tentacles splayed and suckers turned to the sky. A Libyan-dark butcher offered grinning pigs' heads and rabbits skinned but for their fluffy socks. The Saracenic olive vendor had the smooth touch of the harem, her hands all day in oil. A paper-pale bird man perched on the pavement, holding a goldfinch in each white palm, popping them into brown bags for his customers.

We pushed through the disparate crowd, Greek to a man, towards an ironmonger's stand. He sold nail clippers, alarm clocks and sheep shears. His hinges were suited to shed doors not aircraft tails. The only wire he stocked was chicken wire.

'Try by the bus station,' he suggested when I showed him my single bolt.

We cut through 'Turkish' Splantzia, with its cobbled lanes and carved wooden *haïrti* balconies, beneath a minaret and the remains of a mosque, once part of a monastery and now a knitting factory, towards 'Athenian' Hania. The new town was a congested clutter of cement and neon, of shoe shops, 'donat' bakeries and *souvláki* kiosks. We strode between farting mopeds and into the Alpha Bank to cash a cheque. A fishmonger sold fresh squid up and down the queue inside the building. Behind the counter the manager's children – off school for the day – broke the foreign exchange computer.

On frenetic Kidhonias Street I found hardware heaven. Here were glaziers, ship's chandlers, plumbing stores and paint outlets. One shop sold only garden strimmers, which hung by their necks like garrotted turkeys. Another specialized in plastic pipe: blue, yellow and translucent serpents curled along the pavement.

A sun-baked shopkeeper read a text message outside his kitchen fittings showroom. I produced my bolt and he directed us to a dark doorway. The only light inside trickled through the greasy street window. In the dim, far corner a young man yawned and thought of lunch.

'I need to find a bolt like this,' I said.

'Sorry,' he said, not looking up from polishing his scales.

Around us there were row on row of little, numbered cardboard and wooden boxes, reaching from floor to ceiling, stretching back into the recesses of the shop.

'No?' I asked.

'Sorry. Nothing.'

'But here,' I said, reaching up to open a box, 'this is the right size.' I dug into a container and pulled out an identical stove black 6 mm bolt. He stocked a cornucopia of lengths and threads: M8, M10, square and CSK heads, hardened steel and zinc. 'These are perfect.'

'These all I have,' he conceded. Idle. Uninterested. 'But I have nothing like your bolt.'

I didn't understand him. It was a perfect match. The problem didn't seem to be my poor Greek. It was that he didn't know me. Or my family. He simply couldn't be bothered to help.

'Do you have any screws?' I asked.

I needed long, thin self-taping wood screws for the aircraft's tail so it wouldn't disintegrate in flight.

'Not many,' he told me.

There must have been five hundred varieties of screws on his shelves. I helped myself, pulling down box after box. The shopkeeper paid me no heed. He rang his wife to say he'd be home in ten minutes. I asked him for locking nylon nuts and washers. He had them too.

'What are you building?' he eventually asked, lifting his heavy lids while pricing the items by weight on his scale.

'An aeroplane.'

He nodded with indifference, though it was clear that word had spread even to Hania's drowsy alcoves. 'Have you found the runway yet?'

'There's almost too much light today,' said Ariadne, my miniaturist engineer. 'I have to close the blinds.'

'Lock the doors too,' moaned her son Apostoli. 'It is too early to meet the daytime.' And his jealous girlfriend.

Katrin and I had walked down streets named after heroic socialists to find their building, a sixties block of four flats shared between relatives. Its kitchen doors opened onto the road and mulberry trees which stained the pavement. We found two chairs unoccupied by nodding figures or hydraulics schematics, pushed back the empty coffee cups and full ashtrays and sat down.

Ariadne had just arrived home from the base for lunch. Brisk. Overworked. Apostoli was still waking up.

'A good evening?' I asked him.

'Too good,' he replied, cradling his head in his hands. 'Please ask my mother to make me a coffee how I like it. With three bubbles.'

Ariadne ignored him, sucking on her cigarette, producing leftover fruit salad for us. They appeared not to be speaking to each other.

'We have two problems,' she said, dishing out segments of apple and orange. 'First, my nine-times-wise son's engine is too small. Like his ambition.'

'I am sorry,' said Apostoli without lifting his head. 'My mother says the water pump motor hasn't enough power for your aircraft.'

'I spent a morning testing it on the bench. It isn't even two horsepower.'

I told them that a villager had offered me the engine from his old Minotaur.

'A *mechaní*?' despaired Apostoli. 'You cannot fly with a *mechaní* engine.'

'In aircraft design the engine is always chosen first not last,' said Ariadne. 'You are trying to reinvent the wheel.'

'Which is sort of what I'm writing about.'

'What's the second problem?' asked Katrin. As if not having an engine wasn't enough.

'The second problem is that Maleme is impossible. There are many laws about flying in Greece and to use a military airport needs permission from the Ministry of Defence.'

Which wasn't a surprise.

'It is impossible to do this flight legally,' groaned Apostoli.

'There is also a last problem,' continued Ariadne.

Which made three.

'You can break your neck from a height of three metres.'

'That we know,' said Katrin, reining in her fears.

'I don't want to kill myself,' I reassured them. Which was different from not caring if I lived. 'I'll be happy with one Wright brothers' hop.'

The Wright brothers' first attempt at powered flight had ended in a crash from an altitude of perhaps fifteen feet. On their successful second flight the Flyer had hopped 120 feet.

'You will need to do many little bumps along the runway to learn how to fly your aircraft,' said Ariadne. 'Apply power, a little stick, hop up and set back down.'

I wondered aloud how Yióryio's early aviator had learnt to fly.

'Yióryio's aviator?' asked Ariadne.

I told them the villagers' whimsical story.

'Yióryio said he flew when?' she said, amazed by such rural nonsense. 'Come oooon. No one flew in Crete before 1935.'

'Anyone can come up with a myth after enough *raki*,' suggested Apostoli.

'I'd like to look into it,' I said. I believed that embedded in most legends was a kernel of history and this myth – like its ancient precedents – gave me a sense of continuity, not by providing facts and reason, but by helping me to live with uncertainty.

'Don't waste your time. The priority now is to teach you how to handle an aeroplane.'

'So not to carry you home in a wooden box,' added Apostoli helpfully.

We followed him into his bedroom. In the corner beneath a poster of an F-14B Tomcat was a home computer. Apostoli tried to load the flight simulation disc. It didn't work. He'd forgotten to switch on the hard drive. 'I'm sorry but I have not had my coffee,' he grumbled.

'This programme is very realistic,' Ariadne told me, taking charge, elbowing him aside.

'I learnt to fly on it,' said Apostoli, 'and the Sopwith Camel is closest to your aircraft.' He selected the biplane and Maleme airport options. 'This is the throttle.'

She handed me the joystick. 'Now you are in control.'

I crashed on take-off.

'Maybe a Cessna will be easier,' Ariadne said, typing at the keyboard. 'And you can try Hania airport where the runway is longer.'

I crashed again.

For two hours I tried to fly, ploughing off the runway, stalling on take-off, spinning into the control tower and losing wing after wing. The control stick shuddered. The sound effects cracked and boomed. When at last I managed to get myself airborne I couldn't maintain level flight. I slammed into the lighthouse at the entrance to Hania harbour. I hit a flock of seagulls above Galatas and plunged to earth. It wasn't that flying was hard, only that crashing was so easy.

'We have lots of aircraft,' said Apostoli, awake now and not laughing. 'But you have only one Woodhopper.'

'And one life,' Ariadne said, restarting the programme. 'Remember this is not a game.'

Katrin had to leave and go for a walk, unable to watch in her anguished state.

By the end of the afternoon I had managed three bumpy

landings but my shoulders and back were tense, my arm exhausted. I was also scared, which I had never been in a real aircraft. For the first time in my life I was frightened of flying.

'I think that you should do this again,' Apostoli suggested.

Ariadne too was worried.

'I'm thinking that maybe you go straight to some sort of accident,' she said. 'Out of eight landings you crashed five times. That means you are five times killed. I don't know if you have a sense of flying.'

'Let me try again,' I asked.

I remembered reading once about the contrariness of flight, that novice pilots reacted by instinct, for example reducing speed or altitude when in trouble, and that these instincts were wrong. This suggested to me – as all pilots started as novices – that new instincts could be learnt. Flying wasn't a matter of intuitive reflex but rather of adapting old senses to be used in a new way. Rather like learning to ride a bicycle. Or beginning to live with death.

It was a question of practice.

We skirted the old harbour under a full moon, avoiding the restaurant touts who had appeared like fireflies with the warm weather. In her desire not to disappoint a guest, even one who appeared to be on the verge of suicide, Ariadne had offered to introduce me to the town's archivist. I was determined to get to the bottom of Yióryio's story.

The Mosque of the Jannisari was the oldest Ottoman building in Crete, erected in 1645 when the Turks captured Hania. It hadn't served as a mosque since the Muslim population were deported to Turkey in 1923. Its onion domes brought to mind a miniature nuclear power station, crackling with capricious energy.

Inside the building was decorated for a photographic

exhibition like a turn-of-the-century bordello, decked in peacock feathers and boas, with woven red carpets on the walls and plastic orchids in cellophane. In front of a long, gilded mirror, a clutter of candles oozed rainbow-coloured rills of wax. In the centre of the room the cream of Hania society sat in armchairs eating almonds, drinking Fanta and leaping up to smudge their lipstick on newcomers' cheeks. They all looked a little odd, and rather inbred.

'Welcome to my opening,' gushed Roxanie, stepping forward to embrace us, leaving a powdery print on my spectacles.

Our hostess wore a black jumper with a bunch of silver pastiche flowers pasted across her ample chest. Her round fruit face was framed by electric black curls. Ariadne introduced us to her and she held out a stubby hand. I wondered if I was expected to kiss it.

'I'd not expected to see work like this here,' I managed to say, looking around at her display.

Roxanie smiled. 'It is a pleasure to meet a connoisseur of art.'

I had thought that the evening's photography show would be a collection of period prints, given that Roxanie was the director of the Civic Archives. Instead, in between the boas there were a dozen misframed blow-ups of sunrises, kittens and fireworks, many taken with 'trick' optical filters.

'It is my hobby,' she told us. 'I make the time to take photographs when I'm out and about, as I so often am. My camera is always in the glove box. *Kalí spéra*, Dimitri!'

Roxanie shoved away other guests to kiss the marble-smooth deputy mayor who had swept into the gallery dressed in a silk waistcoat. Her total lack of embarrassment impressed me. Glowing under his attention she seemed somehow stretched, as if from too much sun or ambition.

'In Hania if you want to do something you just do it,' explained Ariadne. 'You don't worry what other people will think. I like this woman because she follows her own fashion.'

A group of bemused tourists gawked through the open door.

'At her last exhibition there was a bomb scare,' she went on. 'Every guest had to be searched as the Culture Minister was guest of honour. Many guns and knives were found, which is normal in Crete, and the police wanted to make arrests. Roxanie told them, "I am responsible here. Not you. Arrest them if you want but not at my exhibition."'

When Roxanie slipped back into our orbit, Ariadne told her that I was a writer – 'an important writer' – researching a book on the history of early flight in Greece.

'I knew you were a man of taste,' dripped Roxanie. 'Did Ariadne tell you that I am a poet too? I have written the history of Crete in a poem, and represented Greece in a UNICEF competition.'

'I was told that the first flight took place in Crete around 1910,' I said to her.

'It is true,' interrupted Dimitri, the politician, sipping his Fanta, 'though not *from* Crete but *to* Crete. The pilot flew from Athens to Kolimbari. He landed in the sand but couldn't save the plane so left it and married a local girl. My wife knows the story.'

'My *mother* knew the story,' said his bean-thin wife. 'But it was in 1935. A group of Italian journalists flew here to report on Venizélos' uprising.'

'The uprising was in 1905,' corrected Roxanie, muscling back into the centre of attention. 'And I know for a fact that the journalists were French.'

'French or Italian, it doesn't matter,' dismissed the wife, 'but the pilot married Maria Vrontulakis' grandmother.'

'I thought she married a Communist,' said the politician.

In an attempt to keep them on the right subject I said, 'I can't believe that the first aircraft came here as late as 1935.'

'My wife watches too much television,' explained the politician, intending to put me right. 'Roxanie told us that the uprising was in 1905.'

'That's too early for Greece,' I said. 'Blériot hadn't even flown the Channel then.'

'I don't know your Blériot,' he huffed, national pride slighted, 'but Daedalus and Icarus were Greek. Is that early enough for you?'

Ariadne came to my rescue, saying now that I was writing a travel article and asking Roxanie if she could guide me through the archives. 'No one knows better the history of Hania,' she added.

'I just want to know the true story,' I repeated.

'I love this island with my life,' disclosed Roxanie.

'Maybe you can help him?' asked Ariadne.

'Do you know what people call me?' she asked me. I didn't dare to guess. 'The Tigress. Because I get what I want. It is better that everyone fears you a little rather than stabs you in the back. Come see me in the morning.'

With a wink she turned back to her guests.

'Have another Fanta,' said Ariadne.

Cretans told me what they thought I wanted to hear. As I'd learnt, a good story was valued above hard facts, especially if its telling enhanced the status of the storyteller. Guests were to be pleased so details were always exaggerated and promises made. Which created difficulties for historical research as well as aircraft building.

I began to suspect that there were two types of truth in Crete: first, the elusive truth of objective fact and, second, the common, 'ethnic' truth where the facts were mangled to suit the storyteller and his beliefs.

For example, visitors to Cretan museums assumed that the Turks had made no contribution to the island's history. After all there were few artefacts displayed from their centuries of occupation. Visitors would read nothing about the accusation that Greek archaeologists had, since the nationalist Joseph Hazzidakis took over the Syllogos in 1883, thrown away everything Ottoman in an attempt to ethnically cleanse the past.

In popular culture the expulsion of the Muslims was portrayed as the reassertion of the native society. In fact the deportees had been Cretan, descendants of the mass apostasies of the eighteenth century. The ancients may have been of pure stock – Minoans, Mycenaeans and the rest – but modern Greeks are a mixed race, as are most Arabs and Europeans. A Haniot on the *Flying Dolphin* hydrofoil probably had more Venetian or Ottoman blood than Dorian. To deny this was to distort the facts, to proffer an 'ethnic' history for nationalistic ends.

On the television that evening there was a report on the opening of the new Athens airport. It had been arranged that Olympic would be the first carrier to land there. Except the flight was delayed and arrived late, after a punctual KLM aircraft from Amsterdam. But the national news did not carry pictures of the KLM flight. Only the Olympic aircraft was shown, thereby celebrating national pride, even though everyone in Hania knew the true story – and was content to let the two truths coexist.

'Ethnic' truth lauded Greek egotism, uniting a richly diverse people by declaring that nothing divided them. It didn't make my search for the truth clearcut. At least the hard facts in the archives would put me straight on the matter of the first pilot.

★

'I am a patriot,' Roxanie declared the next morning over the top of a foot-high mound of local newspapers. 'If I could live again I'd live in 1905.'

'Why then?' I asked.

'Because I have flame. Because I like guns and want to fight for my island.'

Roxanie was no bookish, dusty archivist. There were half a dozen firearms in her office: an ancient double-barrelled shotgun in the corner, two long flintlocks in a glass case and a drawer full of revolvers. A stuffed tiger mascot crouched atop a teetering pile of unopened letters.

'If we are invaded again I will fight because first and above all Cretans are revolutionaries.'

Her mobile phone played the Greek national anthem and she picked it off the paper mountain to answer it. In a second the excited tiger was giggling like a schoolgirl.

'It's the former mayor of Heraklion,' she whispered at me, her hand over the mouthpiece, 'congratulating me on my photographs.'

Roxanie had exchanged her floral jumper for a violet velvet dress embroidered with golden butterflies. Her tiny, box-room office was cluttered with filing cabinets, death registers and bunches of dried flowers. Stacks of birth certificates tied up with string were heaped against a blaring stereo. In the exhibition room next door a legless display cabinet was balanced on unopened boxes of cuttings.

'Let me take you through the history,' she offered, after cooing goodbye to the mayor. 'The Cretan State existed from 1898 to 1913.'

I had explained to her that I knew the facts, or at least I had read up on them, but she chose to ignore me.

'During those years we had our own currency, our own police, our own flag.'

It was the period of modern Cretan history which most fascinated me because the island's short-lived independence, before being subsumed by Greek nationalism, ran in parallel with the early, idealistic days of aviation.

'You see the flag?' she asked me.

Above her head a moth-eaten white cross was stitched onto a blue ground with one red quarter.

'The flag was a symbol of the population of Crete,' she explained. 'The blue represented the Greeks, the red was the Turks. But the meaning was understood that in time all the flag would turn blue.'

With the expulsion of the Muslims.

'Union with Greece was wanted by everyone,' she said, which was inaccurate. 'That was the objective of all the uprisings.'

During the nineteenth century the Ottoman Empire had begun to collapse, destabilizing the eastern Mediterranean. By 1832 the Greeks had driven the Turks off the mainland and out of the Peloponnese. But Crete remained under the control of the Sultan despite uprisings in 1833, 1841, 1858, 1866–8, 1878, 1889 and 1896–8. The great European powers, dreading a Balkan war, wished to sustain the status quo in the region. The governments of Britain and France propped up the Ottoman Empire while popular campaigns across Europe armed the Cretan mercenaries. One British consul wrote, Crete 'is a place of insurrections, murders, depredations, devastations of olive yards and vineyards; an isle of unrest and a witches' cauldron'.

Then, in 1897, a Greek expeditionary force landed near Hania and attempted to annex the island. To avert a greater conflict, the Great Powers were impelled to occupy Crete with an international force of British, French, Russians and Italians. The following year a mob of Muslims stormed through Heraklion slaughtering Christians, including seventeen British soldiers and the British Vice-Consul. The European powers responded by sending squadrons of ships, evicting the Turkish troops and establishing a nebulous and ill-defined national government.

The Cretan State was a temporary compromise. Its governor was a frank and charming Greek prince, the favourite nephew of Edward VII and Queen Alexandra. The role of the International Occupation was 'to protect the flags of the Protecting Powers and the Suzerain of Turkey from the nationalistic and patriotic ardour of the Cretan'. It was also to safeguard the Muslim civilians from their Christian countrymen. Despite bloody massacres and decades of massive emigration, about thirty thousand Muslims, or 15 per cent of the population, remained on the island.

The Ottomans – whose symbolic sovereignty, in the words of Balfour, 'found expression in a flagstaff at Souda Bay' – had left Crete in a state of disrepair. There were no good roads and a great need of bridges. Prince George, the twenty-nine-year-old sailor-cum-governor, had no political experience. He could find no competent person to take charge of the island's finances and, again according to Balfour, 'had to look into everything himself – including the supply of boots to the gendarmerie'. One bridge which was built swallowed up the whole public works budget for the year.

Yet Nikos Kazantzakis wrote in *Zorba the Greek*, 'Lord, let my paradise be a Crete decked with myrtle and flags and let the minute when Prince George set foot on Cretan soil last for centuries.' Most Cretans embraced the prince who – with the Royal Italian Carabinieri – managed to disarm the insurgents, taking in eighty thousand rifles from a population who considered a gun as a member of the family. But George never accepted that he was only a figurehead, used by the Great Powers to stave off a greater war.

'It was a time of fantastic dynamism,' enthused Roxanie. 'No one could expect freedom after so many years of slavery. We held an International Fair in Hania. It lost money, of course, but it was a beginning. We wanted to be a part of the world. To make up the time lost.'

Prince George was governor for eight years, undermined in the end by ambitious politicians, insurrection and his inability to distance himself from events. In 1906 three thousand royalists protested against his departure, staging a gun battle in the streets of Hania. The prince had to be spirited off the island under the cover of darkness to prevent him from being carried into the mountains by his supporters. He was succeeded by Zaimis, a former Greek prime minister, whose calm, studied impartiality offended all sides. So as the Cretans had rebelled against the Venetians and the Turks, they now rebelled against the international commission designed to ease them through independence. When Zaimis was away from Hania in 1908, the Cretan flag was pulled down and a Greek one raised in its place. In response an attachment from the international fleet came ashore and cut down the flagpole.

'Those years were really Crete's golden era,' said Roxanie. 'But everyone still wanted union, even if it meant becoming part of a small country with the flaws of a great country.'

In 1913 the Cretan flag, along with those of the Great Powers and the lone Turkish standard at Souda Bay, were lowered. King Constantine and Eleftheríos Venizélos, the zealous, ambitious Cretan who became premier of Greece, unfurled the Greek flag above Hania harbour.

'We say that we had one hundred revolutions,' stated the

archivist. 'Maybe it is true, maybe not, but what is certain is that we *earned* our liberty. Now if anyone tries to take it away they will have to face us.'

Her telephone rang again and she yelled into it, then kept yelling for five minutes. Her eyes opened so wide that they looked like shelled quail's eggs. But the call wasn't news of a Turkish invasion. It was only the building manager. In top volume she complained about her rheumatic neck while gesturing at a damp wall. On it an 1899 lithograph depicted Queen Victoria, the Russian czar and the French president Félix Faure helping a tragic, tousled Cretan maiden to her feet. Beside them a bearded insurgent and King Umberto of Italy broke the chain of oppression. The ground was littered with skulls, shackles and a trampled Turkish flag.

'The manager didn't want to fix the wall but I convinced him to do it,' she said on hanging up the telephone. 'You see why I am called The Tigress.'

Roxanie levered herself out from behind the desk and rummaged in her filing cabinets. In amongst account books and title deeds she found half a dozen mismatched albums.

'These are the photographs of Hania during those years,' she told me, laying them on the table.

The postcards – for there were no photographs – were of Turkish cafés, busy quaysides and '*petits arabes*' children. The 27th Royal Inniskilling Fusiliers stood to attention to welcome the High Commissioner, and Italian soldiers in plumed safari hats lounged outside a whitewashed hostel. Hania appeared to have been a cosmopolitan outpost and the mix of peoples and tongues captured my imagination.

I reminded Roxanie of my interest in early aviation. When she looked blank I added, 'The first aeroplane.'

'I don't know any aeroplanes,' she said, irritated at being distracted. Instead she showed me a card of the Cretan police band playing on the promenade while ironclad cruisers guarded the harbour entrance.

Not wanting to be diverted by her agenda I asked her how many foreigners – apart from the soldiers – had been in Hania around 1908.

'None,' she said, unexpectedly.

I held up an elegant '*Souvenir de Crète*' postcard written in French. I knew that Prince George's mother had been Danish. I'd read that he had come ashore with two adjutants and a private secretary. Could they have been northern Europeans with a passion for powered flight?

'Maybe some foreign merchants lived here,' Roxanie admitted, reluctant to people her liberated island with outsiders. 'Plus of course prostitutes to entertain the troops. And Jews.'

Hania's Jews were considered to be aliens even though they had lived on the island since Venetian times.

'But it was a Cretan time,' she impressed upon me, preserving an ideal Crete for herself. 'The first time since 67 BC that our island was Greek again.'

I wanted to people Yióryio's myth and so insisted that professional civilians must have been attached to the armies. A Scottish engineer perhaps? Or a French surveyor who'd seen Santos-Dumont fly above Paris?

'Maybe,' she conceded, not wishing to disappoint me, 'or maybe not.'

'And could any of them have remained here after the Great Powers withdrew?'

'All would have left,' she insisted. 'Without any doubt.' But she saw my disbelief and added, 'Apart from those who stayed on and married beautiful Cretan girls.'

Hania lent itself to imaginings, like a spreading mosaic of great age, of forgotten origin, to which pieces had been added over the years. I circled it, considered it and with my back to the sea found a worn, stone seat in the Firkas garrison wall. I gazed across the half-moon harbour. Venetian sailors would have sat in the same seat, watching their galleys being dragged into the vast *arsenali* dry docks, thinking perhaps of wives left at home or lovers in the town. Later Turkish settlers had idled away similar afternoons in this seat, before stopping at a café to smoke a *nargila* water pipe and – at precisely five o'clock (or thereabouts) – listening to the town band murder various waltzes. Their Hania was lazy and rundown, its vizier dozing away the hottest hours, its stark white buildings cracked from earthquakes and patched with plaster. The town's single, tired Pension Tu Cosmos – Hotel of the World – offered one small room, and only travellers without friends ever had the misfortune to sleep in it.

Next, in 1898, Greek merchants, dressed in black Sunday best and waving their homburgs, had stood on the seat to get a better view of the royal yacht decked in flags. Prince George had been rowed towards the packed shore by Cretans wearing wide, blue baggy trousers and fezzes with long tassels. A whole people had, according to Kazantzakis, 'gone mad because they'd seen their liberty'. In the crowd men fired rounds from their guns and threw gold coins in the air. At the official reception the French Admiral Pottier wept.

From the worn stone seat I tried to picture my imagined, would-be aviator coming ashore then, in a second skiff loaded with portmanteaux and a leather case of maps, or disembarking a few days later from the ironclad HMS *Illustrious*, flagship of the Great Powers' *stationnaires* which kept watch beyond the harbour wall.

Or maybe he didn't set foot on Crete until five or six years later, arriving with Esmé Howard, the new British Consul-General. Together they were greeted by the Consulate's two, kilted Albanian cavasses 'in white fustanellas, red sashes complete with silver-handled pistol and daggers and velvet waistcoats embroidered in gold'. Under his arm my aviator may have carried ashore a copy of Lee's Greek myths. Or the *Illustrated London News* report on the first flight of the Wright brothers. Or Santos-Dumont's prospectus for the Demoiselle. He may have seen the harbour, the town and the high white mountains beyond and thought, 'Daedalus and Icarus flew here.'

I'd never know exactly how it began, of course. Roxanie had confirmed Ariadne's assertion that there was no record of any aircraft on Crete before 1935. And in none of her papers could we find confirmation of the flying machine which had crashed in the sand at Kolimbari. As for Anissari, I had only Yióryio and the villagers' fanciful assurances. But whatever the truth, the facts had become less important now, for in the imagined arrival of my aviator there was the reassurance of a new myth.

14. Heavy Metal

'Our greatest blessings,' said Socrates in the *Phaedrus*, 'come to us by way of madness.'

'Good heavens, it's going to be a large aeroplane,' bellowed the Englishman, full of holiday voice and an early glass of wine. He and his wife had caught sight of the garage from the *kafeneion* and found me working alone. 'I had a Tiger Moth in Africa after the war but the wings weren't this big.'

He wore a tired blue blazer despite the gathering heat. She had a peach cardigan and sensible shoes.

'Peter used to encourage me by saying that if we had to come down it would be quite all right,' his wife recalled, 'because several Tiger Moths had landed in tree tops. Didn't cheer me up at all.'

'You're a pilot, are you?' he asked me as he studied the plans.

'I'm learning to fly on a simulator.'

'Bravo, so you know all about it.'

There was no cynicism in his voice.

'I don't have a licence,' I admitted.

'You don't need a licence,' he barked, 'especially in Greece. What absolute rubbish. If you want to kill yourself, why not? It's nobody else's business.'

It was also Katrin's business, I realized.

'You'll have a thoroughly marvellous time,' added the old man. Then he hesitated at the wings. 'But there don't seem to be any ailerons.'

'No.'

Ailerons made it possible to turn in flight and every aeroplane had them. Or almost every aircraft.

'So how do you manage that? How do you turn?'

'I don't understand much about aerodynamics,' I told him.

'Shift your weight around a bit?' suggested the Englishman.

'As I understand it I just use the rudder and wait. I read that it'll be rather slow in responding.'

'Like molasses,' he said.

I showed them a photograph from the original article of a Woodhopper in flight.

'And is the pilot still alive?' asked the woman with the genteel courtesy of the drawing room.

Katrin and I had returned to Anissari with the hardware and settled back into work. It had taken me a morning to puzzle out the parts of the wheeled undercarriage which would frame the pilot. The thin, aluminium tubes had been cut to length yet when assembled on the table they looked like a Cyclops' game of spillikins. I tried to arrange them but they rolled across the counter top, knocked over a stack of brackets and clanged onto the concrete floor. The noise disturbed Leftéri's airport game. He picked one up which had clattered across his runway.

'These are as useless as a girl,' he said, unimpressed.

I liked working with wood. Wood was malleable, could be bowed, carved and shaped. It flexed under pressure and was forgiving of my mistakes, most of which could be remedied with wooden matchsticks and a liberal application of glue. Metal, on the other hand, was cold and unforgiving. It tolerated no margin of error.

I realized that nothing was more important for a successful flight than the correct angle of engine and wings. If I drilled a wrong hole the tubes and boom would be ruined. So I decided to make a test rig.

I bolted a three-foot spruce offcut to the garage wall. Then, with two rolls of duct tape and a ball of twine from Aphrodite, I roughed together the aluminium substructure and hung it from the rig. Next I cut the angled seat back from a scrap of plywood and balanced it against the back of a chair. Its shape brought to mind a coffin lid.

I took the control stick in my hand, with its rubber hand grip borrowed from Aphrodite's Zimmer frame, and stepped into the undercarriage, lowering myself onto the seat. I looked ahead and felt a rush of fear.

Now it had happened twice.

I'd begun to worry about losing my life.

I steadied myself, looked around the garage and tried to imagine the wings above my head. A sudden sense of elation pushed aside the fear. In my mind's eye I took off through the flaking concrete walls to soar high into the deep blue of the Cretan spring. I felt the warmth of the sun, the whip of the wind, the reach towards a borderless, clear infinity. The front of the air frame was no more than six inches from my nose. There was – or would be – nothing else between me and space. I looked down again, and straight ahead, and saw not wood shavings and twists of metal but the rocky earth far below. I swooped out over the sea, a free man again in my heart, and felt the soles of my feet tingle with the thrill of it.

I wondered if Icarus had thought about his imprisonment as he popped on his feathers and wax wings? 'I'm free! I can fly!'

Then the fall.

I'd stopped myself from drilling cockeyed holes in my boom because I'd begun to worry. I'd built a test rig to get it right, so my white legs wouldn't disappear, as Auden wrote of Icarus', 'into the green Water'. I wanted to fly on, to live on, like his father Daedalus.

Mind you, anyone looking at my day's work would have thought that I was bent on self-annihilation. It looked like a dog's breakfast. Yióryio watched in silence as I tried to punch the new holes in the round tubing, the drill bit skating across the gleaming surfaces.

'Don't you have a vice?' he asked me.

I'd left mine in England, bolted to its bench.

'Come with me.'

We drove to his friend's house. The friend's mother was digging onions in the garden and did not meet our eyes. In a shed behind sacks of chicken feed was a workbench and vice. We dug them out, disturbing spiders and a scorpion, and loaded them onto the back of Yióryio's pick-up.

'How many years have you?' Yióryio asked me as we turned back to the garage. 'My friend had thirty-one when he die. He crash motorcycle last Christmas Day.'

Each morning I would wake with a precious dash more joy for living. My spirits were lifted by the momentum of the day: by eating breakfast beneath mountains, by walking in the Cretan sunshine, by listening to Yióryio's stories in this place so full of life.

But come afternoon my feeble optimism lay under siege. My mother's telephone number echoed in my head, repeating itself over and over like a broken recording. I stared at the watch she had given me as a schoolboy. It had stopped working twenty years ago, and began running again the week after her death. I imagined her choosing it, picturing it on my wrist, buying me the timepiece which would outlast her. Once or twice alone in the garage I even called out her name.

By evening I'd catch my tired reflection in the *kafeneion* mirror. I'd wish that the construction was completed, that I had already made the flight. I would try to let go of my ambition. I knew that I didn't have to bring myself close to death to appreciate the value of life. I could take the sensible option, walk away from the garage and invent the rest of the story. I was neither pilot nor engineer. I was a grieving, bull-headed writer using words to hold onto that which I could not bear to lose. I needed to make a journey from a kind of prosaic madness back to imaginative sanity, from being unhinged to being reconnected to the world. The writing of the story would be my true flight, my release, the soaring of imagination. I was filled with doubt, confused and strained. I'd ask Yióryio for another drink.

In Greece one can feel closer to the stars, especially after eight glasses of *raki*. In the dark I carried our rubbish up the road towards the village wheelie bin. A breeze had blown up and the groves of olive trees were brushed and harried by it. As the air rushed past my face, I lifted my arms up towards the night sky, towards the points of light which pierced the black canopy. I wanted to swoop up into space, where there was only air and lightness and no matter.

I reached the bin and opened its metal lid. Inside a dead sheep lay on its back, hooves and tits up. I dropped my rubbish into the stink, slammed shut the lid and turned back to face the dark.

★

Overnight the high, harsh east wind gathered strength. By daybreak it ripped over the mountain tops, shredding tendrils of cloud on the sharp peaks, then surged down the valley to fan out over the sea, pressing blossom patterns onto the flattened surface.

I was all but blown off my feet on my walk to the garage. The building moaned and rattled as I clamped the tubing in the vice between lengths of timber, sticking masking tape at the point to be drilled to stop slippage. While shelves of heavy cloud raced across the plateau I calculated angles and bolted the A-frame substructure to the boom. The boom was held into position by the coffin-lid seat and the first pair of bracing wires. Above it rose the kingpost, a metre-long tube from which the web of supporting wires would run to nose, wings and tail.

Outside the mad storm splattered the village with tepid rain then baked it under a blistering sun. Apostoli sent a text message. 'God is shaking his sieve. I cannot come today.'

At lunch time the wind veered to the south. A dirty yellow shadow swelled over the mountain pass. It tumbled down the green hillsides like a harbinger of doom.

'Yióryio says it's called the *lévas*,' Katrin yelled, running down from the house.

The *lévas* rose in the Libyan desert, whipping up African sand, sweeping it over the sea to Europe. It wasn't a seasonal wind like the *meltémi* and could come at any time of the year.

In minutes a sulphur yellow shroud smothered Anissari, blotting out the sky. Dust devils twisted across the *platía*, stealing around the door and into our eyes. The light leached away and the birds fell silent. Sheep stopped bleating in the hills. All sound became muffled. A *mechaní* seemed to float by the garage. Sophia called Leftéri home, opening her mouth like a fish, but no voice reached us.

Then the sky opened and it began to rain mud. A hot,

yellow stain smeared over the black streets and silver olive trees. Dirty streams jiggered down the valley, into gutters and alongside irrigation pipes. The houses newly whitewashed for Easter were spattered and soiled. Kóstas the policeman sheltered in our doorway for a moment then ran on, desert ochre teardrops streaking his uniform.

We closed the garage and belted through the dim, saturated air for home, expecting earthquakes and tidal waves to befall us. The storm caught us again outside the *kafeneion* and we took cover from its gusts. Ulysses pushed past us into the gale, calling after his gods or father. Polystelios reeled in shouting that trees were down on the Vrysses road. A pair of Yióryio's chairs lifted off the veranda, soared across the road and came to rest in an almond tree.

The first strike of lightning cut through the fine dust, illuminating the world in its dirty brilliance. Teeth-rattling thunder shook the bottles behind the bar and a cloudburst washed away an ice-cream sign. The dust was on the tables and between our teeth. The telephone rang twice then the line went dead. Iánnis didn't look up from his cards as a beachball skidded along the street, blown up the hill from the tourist beaches at Georgioupoli.

By the next morning the storm had passed. Yióryio turned two hoses on the opaque *kafeneion* windows. Polystelios scrubbed his *mechaní*. Greeks always seemed happiest with a hose pipe in their hands. Muddy water streamed off every patio and forecourt, pouring down the high street until the pressure dropped and the supply was cut off.

Above the village the mountains' pristine, white peaks had been tinted sandy brown. In the garage my papers were gritty to the touch. I blew the dust off the wings and they sang under my breath.

Apostoli arrived before I'd unreeled the first length of cable. It was early for him, so early in fact that I thought he might not have gone to bed yet.

'Last night I stayed home,' he explained, 'and flew Concorde from Hania to New York.'

He seemed to spend hours in front of his computer, flying the simulator towards a pilot's licence. He seemed also to have had words with his mother.

'Some of my friends have finished in the army and now they are just killing flies. They say to me, "What to do with life? Find a nice job and then the God has it."' He shook his head. 'I know what to do with life; not waste it. So now we work.'

He was filled with purpose and together we tackled the rigging. We measured and cut half a dozen wires to support the boom and square the cockpit. It was slow work – climbing onto the table to reach the kingpost, twisting the metal rope, setting the tension by hand – but his quiet determination pushed us both forward. He went without his usual frappé, drinking a Coke instead while we made the eye splices.

His only complaint was with my tools, again. He disliked my DIY wire cutters, which weren't hefty enough to make a clean cut in the three-millimetre cable. The sharp, frayed wire ends stabbed our fingertips and tested our tempers.

'I cannot believe that you bring these scissors from England,' he grumbled. 'In Greece we have superior tools.'

He promised to bring a better pair from home so we stopped work at three o'clock, but the next day he arrived empty-handed.

By the end of the third day the wires stretched from axle to tail, kingpost to rudder, sketching in space a figure drawing which described volume. It was as if the plans had been lifted off the blueprints and enchanted into three dimensions.

'Now it comes together,' he said.

With Katrin's help we raised the wings off the floor and balanced them on two pillars of *kafeneion* tables. I aligned the boom and hammered the bolts home. We eased away the stacks of tables and the Woodhopper balanced on its narrow, dragonfly boom.

A breath of wind blew into the garage and she inclined on her boom, tilting in the breeze like a graceful bird tensed for flight. The wings dipped, lifted and in that moment transformed the garage, making it smaller, as the idea of flight became the actuality of a flying machine. I might have had no engine or runway, the half-baked design might be a death-trap, but my aeroplane was beautiful: as light as a Calder mobile, elegant as a Klee painting, a shape at once familiar and original, ageless and modern.

Katrin spoke first, her voice touched by wonder and fear. 'You are going to fly this,' she said.

There had been a stone in my shoe for the last three days. I tried to remove it and discovered it was a splinter. Katrin extracted it with tweezers. It was a one-millimetre strand of aircraft cable. The Woodhopper had become part of me.

15. Bones, or How Aphrodite Lost her Virginity

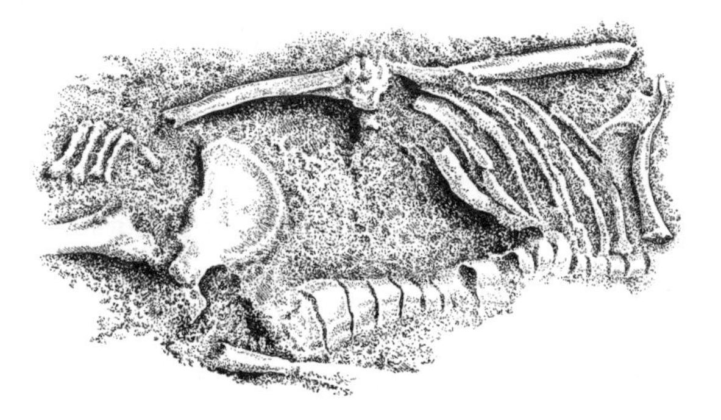

'This is a story about flying,' Aphrodite told me, holding my hand across the dirty kitchen table.

In the months since our arrival every villager seemed to have dug up old tales and adapted them to flight, if only to please me.

'It is a true story of miracles and salvation.'

Apístefto kai ómos alithinó. Unbelievable yet true.

'When I was a young girl four times I dreamt of finding bones – human bones,' she said, 'in our olive grove near Mezaria.'

'Near to the old Turkish cemetery?' I asked. Mezaria was the box of land on the edge of the village where loony Ulysses had fallen to earth. Years ago it had been the Muslim cemetery, until it was unearthed and the gravestones smashed to make a roadbed.

'Not there,' she said, 'but behind the house of Kóstas the policeman. On my grandfather's land.'

Which, of course, was the same place.

'Four times I had the dream,' she repeated, 'but the fourth time I heard a man's voice calling to me, "Come unbury me." I knew the place so I went digging in the field but found nothing.'

Polystelios was outside lashing together the tripods above his spreading runner beans. I had stopped in to buy eggs and, after a glass of wine, he had returned to work. Katrin was waiting at home to cook breakfast.

'Then one night I heard the voice again,' she said, squeezing my hand, not letting go. ' "You are digging under the wrong tree," he told me. The next morning I dug at the right spot and found an icon of Agios Ioannis.' She crossed herself as she repeated the saint's name. 'A silver icon lost in a field; a miracle. I carried it home, polished it until it shone but in the morning it had gone.'

'Gone?'

'It had flown back to the grave,' she insisted, pausing to remind me that this was a story about flying. And that her story was better than the fable of the winged icon of Panayía Kardhiótissa. 'Polystelios was a good man. I persuaded him to go back with me. He planted his spade in the earth, turned it once and we found bones.'

'From the cemetery?' I asked.

'*O po po po*,' said Aphrodite, 'why don't men listen? *Ti na kanoume?*' What can we do? 'They were the bones of a Christian.'

'You could tell this?' I asked.

'Because of the icon,' she reasoned. 'There and then I pledged to build the chapel on the spot and Polystelios promised to help me. But my father was not a wealthy man. At home we lived on *hórta* and *tiganopsomo*.' Wild greens and fried bread dipped in oil. 'I started making crochet, and walked around the valley selling my work. Polystelios sold his donkey. Together we raised enough money to build the chapel to Agios Ioannis ourselves, with our own hands, and placed the bones in a box inside.'

At last Aphrodite let go of my hand to rub her abdomen. She swallowed a mouthful of wine. She didn't like the taste but drank because she thought it was good for her veins. A spider's web of burst blood vessels flushed her face and she looked ill.

'Then Polystelios and I married and I never again dreamed of the unknown man, because his bones now lay in consecrated ground.'

I waded knee-deep through sweet air. Yióryio strode ahead of me, his footfalls kicking up waves of peppermint and oregano. Their perfume rose above the rust-brown soil, out of the low mountain sage and twiggy thyme uneaten by goats.

He was taking me into the hills above the village to meet an old shepherd who might have known the early aviator. We would have driven – no Greek walks when he can drive – if his pick-up hadn't lost its exhaust again. But then I'd have missed the flanks of bright yellow gorse, the fennel with fist-thick stalks and the first hint of magenta blossom on the Judas trees. I gulped in the scent, warmed in the morning sun, and felt as if I was drawing strength from the earth.

'When I am a boy I work with my father in that field,' he said, stepping along the path. 'We have almost nothing to eat: olives, an onion, some bread maybe. In spring we catch eel fish in river. We look and see their holes and we build little pool. We drop in the water *flómos* twigs.' *Flómos* was a low bush with no bark and milky sap. 'The milk choke eel fish. He comes out of holes and we spear and skewer on a stick to cook like *souvláki*. Not any stick but stick made from carob or olive. Oleander twig gives bad taste.'

He kicked at the earth.

'Now I read in Athens a teacher ask her school to draw picture of a chicken. Twenty-nine children draw a supermarket chicken for the pot. Only one draw bird with feathers.'

I couldn't explain his agitation, unless Sophia had overspent at the Co-op. Or Leftéri had expressed a preference for packaged Yoplait instead of his grandmother's fatty home-made yoghurt. Not that the old lady would have minded; it would have given her more time to surf between satellite channels. I'd never seen her without a television remote control in her hand.

'I tell you now that we Greeks are shits.'

'What do you mean?' I asked. I couldn't agree with him now, though the *pilótos* might have in weeks past.

'It used to be every house had an oven. Now who knows how to make bread? Or to plant tomatoes when moon is going full? If a war is coming and supermarket closes, how many will not eat because they cannot skin a lamb? We Greeks give away our culture.'

'You want me to write this in my book?'

'It is truth.'

'And if I do will I ever be welcome in Anissari again?'

'No, we will cut off your balls.'

Which was more in keeping with the Cretan adage *vlépe, ácouse, klíse to stoma*. Look, listen and lock your mouth shut.

Across the valley steps of white houses rose out of the olive and orange groves. Above the newer buildings decrepit, forgotten lanes opened onto vistas of distant sea or closed into dead ends of amnesia and ruin, the shells of past lives propped up by fig trees and old men's stories. It was hard to imagine how many generations of houses had been overbuilt, how often ancient olive beams had been reused, how there could be so many television aerials in such a small village.

'Every second day make the earth into pyramid,' continued Yióryio, advising me on growing potatoes, 'so the stem reaches to the light. You don't know this?'

Beyond the edge of Anissari rose the chalky dome of a chapel, hidden among the trees on the Exopoli road.

'Is that the church built by Aphrodite and Polystelios?' I asked Yióryio.

'In all his life he build only chicken shed,' he scoffed, stopping to laugh. 'It blow over last winter.'

I told him Aphrodite's story – or at least the story Aphrodite had told me – and he laughed again.

'There is only one bone she dreams of,' he said.

'So what really happened?' I asked, watching my footing.

Yióryio leaned against a wall to catch his breath and light a cigarette. He told me that Polystelios had been a feeble boy.

'He grow up thinking ghosts – maybe even Minotaur – live under house.'

Polystelios had been born in the room in which he would live his life. As a child he had stolen bones to throw under the chest whenever his family ate chicken, which wasn't often. At Easter, if his father could afford to butcher a lamb, Polystelios took a shin as his childish offering. After a year his mother could no longer stand the smell and found the dozen stinking vertebrae and wing bones.

'His father beat him for one day and every day for all month.'

Then, instead of a Minotaur, Polystelios convinced himself that God and the saints hovered above his head, close at hand like the gods of Homer, talking to him and taking part in his life.

'One time he took wood to Vrysses to sell for flour. He met a stranger on the road, gave him a ride on the donkey and came home saying he met Agios Antonis himself.'

'Maybe he had.'

'And maybe Zeus strike us down with thunderbolt.'

According to Yióryio there was only one miracle in Polystelios' soft-headed life and that had been Aphrodite's determination to catch him.

Polystelios had never tortured frogs or caught birds by coating branches with glue. He had been a quiet boy, keeping watch over his father's few pear trees, happiest in his own company and that of the all-hearing gods. His upbringing had been strict, though no more so than other families. In those years most children were forbidden to look their parents in the eye. Once he had been playing in front of the house with marbles, or at least little balls of half-baked clay dipped in paint. There hadn't been the money to buy glass marbles. His mother had called him inside for dinner and Polystelios put his damp marbles on the brazier, fuelled by cracked olive stones. As the family ate their meal the clay of the marbles began to steam. Then they exploded, hurling tiny burning paint fragments around the room. One landed on a new cotton bedcover with its fine crochet work almost completed. His father smelt the burning and turned to see the smouldering hole singe away a week of his wife's work. Polystelios was beaten again.

A pannier of mulberry sticks had been broken over the backs of every village boy before his sixteenth year, but only Polystelios was brought to tears.

'He was chicken shit,' Yióryio said fondly of his friend.

Polystelios was no rebel. He never took up his rifle to join an insurrection in the mountains, though not because the Turks had long since departed from the island. He didn't like confrontation and never contradicted his elders, confining his thoughts to his prayers. Only once did he display a flash of defiance, persuading his mother to work on a Sunday and fix a tear in his jacket so he could attend a wedding. But on the walk there his ears filled with the furious beating of angels' wings and his heart pounded with guilt. He turned for home and went to bed for a week, only getting up to burn the jacket in the fire because of its *kiriakokendies*. Sunday stitching.

His beaten-dog compliance attracted ample-hearted Aphrodite, as well as the leaner village girls in their tight-waisted skirts. Like Polystelios, and Ariadne a generation later, Aphrodite had grown up in a large, strict family. In her grandmother's day girls had been banished from their father's sight within a month of their first menstruation. Her own father had treated her as a chattel, asking his wife 'How is your daughter today?' while pressing Aphrodite's brothers to his chest. 'I have five children,' he would brag to strangers at the *kafeneion*, adding as an afterthought, 'plus one daughter.' Again the discipline and prejudice were not unusual, simply the product of a stone-broke, defiant society which had long been oppressed. But they instilled in Aphrodite a desire to protect herself by having a man whom she could mould. She didn't want to marry a high-booted Adonis with silver-handled dagger only to be battered through her adulthood.

Once when picking *hórta* in a field adjoining the orchard she disturbed a snake. She ran into the pear trees, knowing that Polystelios would be there dreaming away the afternoon, but instead of comfort, or even a hesitant first touch, all she got was an earful of piety.

'Don't you know that the snake was Agios Ioannis telling you not to work on his day?' he told her.

Aphrodite sensed that to snare dreamy Polystelios she needed a miracle. But the saints were an independent lot.

They dished out wonders for higher ends, not to satisfy base instincts. Neither St Therapon, who was invoked for all kinds of healing, nor Aphrodite's beautiful namesake, born of the white foam which formed when Kronos threw the genitals of Uranus into the sea, would get off their high chairs to help a love-struck girl. Aphrodite realized that she would have to conjure her own miracle.

'I not know where she found icon,' said Yióryio. 'Maybe she stole. Maybe she bought from clove peddler. He walk from villages selling clove oil by thimble. He always had sack of junk on his back.'

'Or maybe she really did dig it up,' I said, not wanting to disbelieve Aphrodite's version.

'*Nomízeis óti mólis vgíka ap' t'avgó?*' Do you think I just came out of the egg?

Aphrodite buried the icon beyond the orchard in her grandfather's field then stitched together a story for Polystelios about a voice in her dreams. He was curious but unconvinced, until she started digging in the field. In a fine moment Aphrodite scrabbled in the soil, broke a nail and unearthed the tinker's icon.

'It's Agios Ioannis,' said Polystelios in wonder, falling to his knees in recognition of the disciple who had taken the Virgin to Ephesus after the Crucifixion. In his hand St John held a chalice from which a snake issued, the snake which had chased Aphrodite from her *hórta* picking. It was a fortuitous fluke which she hadn't noticed.

'You must take the icon to Papás,' insisted Polystelios. 'It is a sign.'

Aphrodite didn't want to share their discovery with the priest because she feared being found out, but having caught Polystelios' attention she knew of no other way to engage his heart. Once they were married she could begin to dislodge his pious extravagances. Until then she abided by the maxim that men must obey God and women obey men.

She took the icon home, trying to keep it from her parents, but her mother caught sight of it and told her father who relayed the news to the priest. She told each of them her story, and with each telling drew herself deeper into its fiction until it became her truth. There was the dream, the voice and the digging in the soil with Polystelios. Only her friends doubted her, knowing that the icon was a sign only of her longing, but their jealousies were more than offset by the priest's insistence on showing her discovery to the patriarch in the morning.

So it was to Aphrodite's advantage that the icon disappeared that night. It had been in the priest's house, next to his sleeping alcove, and he had awoken with a start to find it gone. There and then the priest declared that the icon had flown back to the field, like the Chained Lady of Panayía Kardhiótissa. As proof he cited Aphrodite's latest dream in which a saint's thigh bone sailed through the air and landed on her lap. The priest organized a gang of men to comb the field. After two days' searching they unearthed no icon, no grave, nothing but fragments of a few ancient pots.

As the men lost interest and attention shifted away from her, Aphrodite really did dream of a voice calling to her, and of Polystelios stretched out on the ground under a pear tree. She woke from fitful sleeps, paced the floor and was beaten back into bed by her mother.

At the end of the fifth day Aphrodite returned to the field, not knowing what she might find. She was light-headed from lack of sleep. As if fated by a miracle, though not a very surprising one, she found Polystelios there, alone, searching the earth.

'You're digging under the wrong tree,' she told him and showed him to the spot revealed to her in her dream. He dug into the soft ground, scooping out the soil which had been turned a dozen times that week and hit a solid object. Together Aphrodite and Polystelios tore at the earth, their hands touching, her head spinning, expecting to find the icon. Instead they pulled out a long, grimy bone.

'Do you know the Dance of Isaiah?' she asked him, shocked by her boldness as they held either end of the bone. He shook his head, knowing full well that the dance referred to the marriage service when bride and groom circled the altar. 'Then I will teach you,' she said and they made love on the spot.

'How do you know this is true?' I asked Yióryio.

'Because I see them, like you and Katrin last night.'

'Last night?'

He chuckled, drew on his cigarette and said, 'My cousins and me were there. It was my cousins who stole icon and planted old sheep bone.'

'It was a sheep bone?' I asked. 'Couldn't Polystelios tell?'

'Maybe he not want to know,' said Yióryio. 'He may be chicken shit but he not idiot. My cousins were jealous that other girls like him so make him marry ugly Aphrodite.'

The story – or at least Yióryio's version of Aphrodite's story – reminded him of an old Greek joke. One day God, feeling in a benevolent mood, went to England and met a man named George. 'You've led a good life, George. I'd like to give you the one thing that your heart desires.'

George replied, 'I would like a castle please, sir.' And so God gave him a castle.

Then God went to France and found Michel. 'Michel, you have led a good life. What does your heart desire?'

'If it pleases you, sir, may I have a vineyard near Bordeaux?'

And God granted him his wish.

Then God turned to Italy, found a good Christian named

Franco and offered him anything he desired. 'An Alfa Romeo Spider, *per favore*, Papa, plus – if it is not too much trouble – a blonde in the passenger seat.'

Then God went to Greece and found Spiros.

'Spiros, you have been a good man,' said God. 'What can I give you?'

Spiros thought for a moment then said, 'My neighbour's goat produces five litres of milk every day.'

'Then I will give you a goat that produces seven litres of milk,' offered God.

'Forget that,' said Spiros, 'I want you to kill his goat.'

Yióryio flicked aside his cigarette.

'And the chapel?' I asked him.

'Polystelios pay for the building with her dowry. He buy his way out of sin, spending everything so he could hear saints again.'

'He stopped hearing them?'

'After they make love. But Aphrodite was happy. She become only voice in his life.'

Humility was not a national virtue, nor was honesty. To concoct stories won admiration, as if proof of a man's superior wit. It was said that anyone could be truthful but to be shrewd was admirable and Greek.

'Aphrodite like a chapel. It give her *timí*, prestige, when she give him no children.' And countered the villagers' poisonous, ubiquitous *fthónos*. Jealousy.

In the village no one knew for certain why the couple remained childless. Their consummation had been witnessed in the field, but then a spell of binding may have been cast over them at the wedding. A jealous girl needed only to make three loose loops in a piece of thread and pull them into a knot during the pronouncement of the blessing 'to put the devil between them'. Or perhaps their infertility was the doing of a Turkish ghost, disturbed by the desecration of the old Muslim cemetery. Or even, as Yióryio hinted, because Aphrodite had subsequently given herself to other men in her desperation to escape barrenness. But whatever the cause, all that Polystelios and Aphrodite touched – conception, shepherding and beekeeping – ended in failure.

We turned our back on the white chapel and lost cemetery and continued up the green olive flanks. Cypress trees rooted in the goat rocks, providing shade both for the birds and ourselves. Yióryio began to call for his uncle.

The old man slept in the village but spent his days in the hills. He rose every morning at six, drank mountain *malotyra* tea and ate home-baked bread then climbed to his land with his sheep.

'He has hundred years,' Yióryio assured me as we walked on into the harsh country near to the sky. 'Or hundred and one. He find snails in rocks for lunch, eat with potato and one glass wine. He is *palikaria*.'

The *palikaria* were originally medieval footsoldiers but the honorific term was now applied to any masculine man, whatever his age.

We found his uncle working a patch of earth, behind a mesh of wire fencing. He wasn't wearing traditional boots, blue breeches and head kerchief but a pair of worn jeans and an Athens Olympics T-shirt. He had a thick-boned skull and didn't look older than sixty.

'What are you doing, uncle?' Yióryio asked him.

'Planting olives as any fool can see.'

'But why are you planting trees?' Implying that the old man would not live to see them bear fruit.

'I plan to piss on a lot of graves,' he replied.

Yióryio had brought a bottle of *tsikoudiá* and we sat on broken-backed chairs in the shade as he poured the little glasses. The defiant old man downed his glass in a gulp.

'So you're the idiot building the aeroplane,' he said to me, waiting for his glass to be refilled. He seemed not to be averse to drinking more than a single glass.

'I tell him about first *pilótos*,' Yióryio explained to his uncle, 'and maybe you remember him.'

'I never heard of him.' He turned to me. 'Are you going to fly this machine?'

'That's my plan.'

He laughed, a sound without joy or compassion, and swallowed a second glass. 'Listen to me,' he said, 'when you crash don't hit anyone. Maybe they'll soften your fall but then you will forever have enemies on Crete.'

16. Mach 0.8

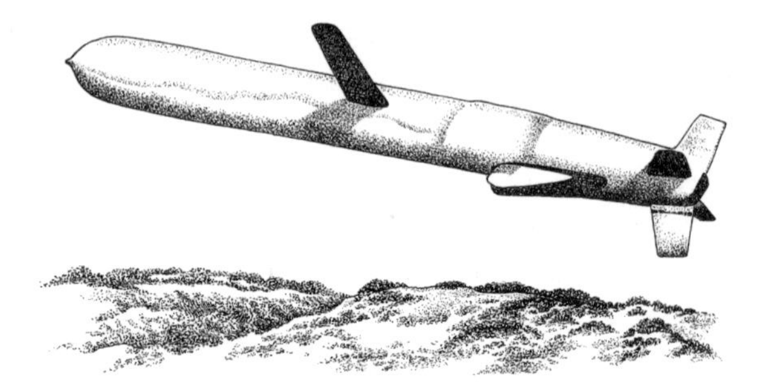

'My mother has found an engine for sure,' said Apostoli over the mobile.

A pool of water gathered around my feet, trickled off the rocks and back into the sea. I had waited three days for his call, carrying my phone from house to garage to *kafeneion*. I'd finally given up in frustration and gone for a swim. The mobile started ringing as soon as I was twenty yards from the shore.

'She has an important friend,' he continued. 'He has engineers who will help in friendship's frame. There is nothing to fear about this.'

Apostoli was excited. I heard voices laughing in the background. 'Are you with the friend now?' I asked.

'I am at the beach,' he replied. 'It is a beautiful day.'

In Crete no course ran smooth. No objective was reached without digression, long arguments and lunch. The extremes of weather – of afternoons too hot to work and balmy evenings too pleasant to waste away on sleep – determined the island's pace. Its history of occupation, defiance and independence set the character. Or so it seemed as I held on for my life, barrelling down a dirt track into a cyclone of dust towards Souda Bay.

Ariadne was behind the wheel, headlights flashing, wipers flipping, driving into the void. The four-lane New National Road was closed and its traffic thundered at full speed along a parallel, pot-holed, residential lane. No diversion had been provided by the police so drivers had made their own alternative route. A school bus tried to pass us on the inside. An oncoming fruit lorry blared its horn. Flying could not be more dangerous.

'When we get there . . .' Ariadne shouted, breaking off to avoid a *kafeneion* table at which two men sat drinking dusty wine. 'When we get there don't tell him where you are building until we find out if he's OK.'

'I thought he was your friend,' I said as a dump truck loomed out of the gloom.

'I've met him once and I liked his face,' she said.

I could have said, 'Your son told me that he was a friend.' Or 'I understood he promised to help us.' But to do so would have offended her and might have led to the withdrawal of her assistance. At sixty miles per hour I found myself humming 'Tiptoe Through The Tulips'.

'We can trust him,' Ariadne assured me. 'I think.'

We swung off the road and into the sunlight, circled the Souda barracks, then bumped across a dry riverbed into a new, labyrinthine business park. The estate looked like a cross between an industrial wasteland and a garish garden centre. Half the units appeared to be unoccupied, their windows boarded and lawns gone to seed, while the other half were decked with bright hanging baskets and riotous pot plants.

Ariadne asked for directions from an ironmonger lounging in a deck chair. He hadn't heard of STN Atlas but pointed us towards the Beyond Fresh ice-cream factory.

'Ignition! Fire!'

We followed our ears for the last 200 yards. A sleek, screaming jet engine – about the size of a bar stool – kicked flickering orange flames out of a workshop door and into the street. Men wearing ear protectors and carrying clipboards huddled around the test platform reading gauges and making notes. At the chief engineer's command the throttle was opened and the jet howled up to full power. I covered my ears with my hands. Behind a curved steel exhaust deflector a pair of smokers puffed on cigarettes and chatted about their vines.

'Our machines travel at four hundred miles per hour,' explained the operations manager five minutes later. On trolleys and suspended above our heads were dozens of pilotless, military aircraft used for battlefield surveillance. 'Except that one,' he said leading us past a Cruise-class missile, 'which flies fifteen feet off the ground at Mach 0.8. Just below the speed of sound.'

My aeroplane would potter at 35 mph, and with luck wouldn't cause collateral damage.

The air reeked of jet exhaust and fibreglass. On the wall there were photographs of Phantoms and Tornadoes. A red sign read – in Greek, English and German – 'Always disable power before servicing stand-off weapons.' I hadn't expected to find a rocket manufacturer in Crete. Yet for all its aeronautical expertise, STN Atlas seemed an unlikely place for me to find a suitable engine, unless I planned to burn up on lift-off.

Until the manager stopped by a metal rack.

'As well as jets we have this,' he said, pulling down a compact, two-stroke petrol engine and laminated wooden propeller. 'It weighs seven kilos, delivers eighteen horse-power . . .' He looked over his shoulder and lowered his voice. '. . . and would suit your aeroplane.'

As he slipped the engine back on the rack Ariadne winked at me. I started to ask the manager about its mounting bracket, assuming that he was Ariadne's once-met friend, but he said, 'Better speak to the boss. He's expecting you now.'

STN Atlas was the Greek division of a European conglomerate of defence systems companies. Its diverse, independent enterprises made Eurofighter flight simulators (Typhoon weapon class), Varo LLC missile launchers, Skyfire-70 rocket systems and SNPE explosive charges 'for Air Forces Insensitive Munitions'. Its air-launched weapons, aviation life-support equipment and landing aids were sold to governments around the world.

We climbed the spiral staircase to the cool executive suite. A secretary batting lanky eyelashes offered me a Coke. Dimitrios Sarris, the CEO, strode across the thick, charcoal carpet to shake my hand.

'Are you a Scot?' he asked in smooth English. 'I studied in Glasgow for two years.'

Sarris was a big, busy man with broad hands and an open face who filled the room with raw energy, rather like one of his jet engines. His hair was close-cropped grey. His heavy polo-neck sweater warded off the air-con chill. On his matt black desk were the business cards of French financiers. On the wall the crests of two dozen European fighter squadrons surrounded the framed plans of a Submarine Spitfire.

'Do you know the Western Isles?' he asked, directing us to chairs, at ease in his command post. 'I love them all.'

We talked about islands and the independence of islanders as the coffee arrived. He told me that he had moved to Crete – 'to this paradise' – because of the Namfi missile firing range beyond Hania on the Akrotiri peninsula. He enthused about the island's mountains, its fishing, its four-kilometre-long runway built by the Americans as an emergency landing site for the Space Shuttle.

With Ariadne's prompting I told him about my Woodhopper and asked him about his engine.

Sarris listened, curious and focused, then leaned back in his chair and crossed his arms.

'I would like to help you but it is impossible.'

I assumed that he was concerned about liability and offered to sign a disclaimer absolving the company of responsibility in the event of an accident.

'Every one of our engines has a serial number,' he explained. 'They can only leave the factory under licence from the Ministry of Defence. And none of them are approved for manned use.'

It hadn't occurred to me that there might be different legal classes of aero engines.

'Then could I purchase an engine and prop?' I asked, again trying to relieve him of responsibility. I hoped that if undamaged he might buy them back after my flight.

'It's not a question of money,' said Sarris. 'I would happily give them to you and, if they were destroyed, it would be no problem. My concern is security. I can only sell to you on receipt of an End User Certificate, signed by your British Department of Trade and the Greek counterpart.'

'And they would have to know that I am building a manned aircraft.'

'There is much bureaucracy,' said Sarris, spreading his hands on the desk, 'because of the fear of terrorism. You understand with this engine you could build a missile and fly into the centre of a city.'

I understood all too well, so tried another tack.

I proposed moving the Woodhopper into his workshop, fitting the engine under his supervision and transporting the aeroplane to an airfield in one of his trucks. I placed myself in his hands.

'Again we need permission even to transfer equipment to our test facility. It is the law,' he told me, ' and I cannot break the law.'

Sarris' refusal was unequivocal. I was grateful for his frankness. The matter was closed.

I thanked him and stood to leave.

Ariadne and the operations manager began speaking at him. Their voices were raised, their argument heated. I couldn't follow the Greek – it was much too fast – but I caught snippets of the debate.

'Can't you just lend it to him?'

'And if he's stopped on the highway?'

'Then he comes up with a story.'

'The police will check the serial number and close the factory for ninety days.'

'No one will stop him.'

'Ninety days and a fifty-million-drachma fine.'

Shouting all the while, though without anger, Sarris called his lawyers in Athens. His contracts department faxed through an End User Certificate. We stood around the desk in a circle, the Greeks yelling while I read the text. The wording of the document was precise.

'There is no way around this,' I said.

'I trust you and want to lend you the engine,' Sarris assured me in English, allaying the Hellenic fervour with the switch of language. 'But I cannot do it. '

'I am a guest in your country,' I said, 'and I am grateful for your trying to help me, but I would never want your generosity to put anyone at risk.'

I shook his hand, thanked him for his time and left him my telephone number. Just in case.

He never called.

Ariadne drove back to town on the slow road.

'He was telling the truth,' she said. 'I like him.'

I didn't feel much like talking. We had found the ideal engine. I had held it in my hands. I couldn't use it.

'If he worked in a Cretan way, he would have given you the engine and dealt with problems only if they occurred. It's too bad he works in a European way.'

She dragged on her cigarette.

'Maybe your book should finish here,' she proposed, 'with an unknown ending. Then the story will symbolize our hope and aspiration. Will it ever get off the ground? Just like life.'

'I need to finish and fly the plane,' I told her.

Three Greek engines – the feeble water-pump, the leaden *mechaní* and now the STN Atlas two-stroke – had proved unworkable. The time had come for me to look beyond Crete. But Sarris' concern about the illegality of my proposed flight now worried me. Even if I did find a foreign engine and the supplier turned a blind eye to the project's unlawfulness, where would I ever find an airfield?

'We've been too obsessed with high-tech equipment,' Ariadne said. 'Remember how the early aviators first flew.'

'They tended to crash and kill themselves,' I reminded her.

She reached over to touch my arm. 'In Crete we say, "Don't worry about anything, we are here."'

'Which means?'

'It means that we are the law and our way is best.'

With that thought in mind we rattled down the dusty, pot-holed lane towards Anissari.

17. Born Again

No car passed the door all morning, apart from a psalter of priests in a black Opel. No bread vans called at the village. No voices reached between the fields. Veils of cloud twisted around the mountains, tumbling downhill to snatch the sombre rhythm of the church bells and lay them across the valley. The weather had turned cool as if in respect for Crete's mourning. It was 'Great' Friday, the day of Christ's funeral in Greece.

We walked the deserted streets, passing abandoned pick-ups parked outside shuttered homes, looking for the villagers. The *kafeneion* was shut. The square was empty. Everyone seemed to be indoors, heavy-eyed and limbed at the end of forty days' fast, though we'd noticed nothing given up for Lent. Except possibly sobriety. Yióryio pushed a window ajar as we passed his house. He looked like death.

'No work today,' he instructed me.

Behind him in the dark a dozen unlit candles – wrapped in

the Greek flag and moulded in the shape of lilies – awaited the return of the light. Sophia had hard-boiled the eggs, dyed them red and was making the sign of the cross with them on the faces and necks of her children.

'Go home,' Yióryio said in a low tone and pulled the window closed.

Since ancient times the Greeks had marked the vernal equinox. The Minoan goddess of nature danced with a golden bough, symbolizing the death of the year and its regeneration. Later Persephone, daughter of Zeus and Demeter, the goddess of fertility, returned from winter exile to restore the earth's fecundity. The Eleusinian festival, onto which some maintain Easter was grafted, extended the metaphor with the promise of immortality. Christ died and, as the ancients before him, was resurrected into life everlasting. Easter was the ageless Radiance, the annual rebirth, an ascent from barren days. And like life itself it was bound to darkness and death.

The morning flitted between showers and damp sunlight. We sat on the balcony watching the breeze stir the olives. Snatches of sound reached us – the tolling of the bells, the chanting of a male choir – only to be whisked away on the wind. We caught the cries of sheep then saw blood stream down the cobblestones of our narrow lane, pooling around tin pots of basil beneath our balcony. Behind the house two sheep lay panting against a whitewashed kerbstone, bleeding to death, their throats cut. Three skinned animals hung from a beam, their eyes blank and teeth set in grimaces. Their fleeces dried in the arms of a lemon tree.

An hour later a flush of village girls appeared in the fields beyond the road. They were dressed in red, scarlet nymphs wearing Nikes, darting between the olives, gathering spring flowers into white tablecloths. They carried narcissi, irises and blithe spirits into the church to decorate Christ's bier.

The bells rang on. The nymphs returned indoors to mothers boiling wheat, sultanas and pomegranate seeds, making *kóllyva,* a food for the dead and offering to the gods. Socrates butchered another lamb in Christ's name, tethering its legs and slitting its throat with a flash of his knife. I watched the animal bleed to death. He cut a notch in the skin of a rear leg then inserted a thin length of bamboo and blew. As his son prodded the lamb's body, the air separated skin from flesh to allow easy skinning.

At lunch time the *kafeneion* stayed closed. Unfamiliar cars with Athenian licence plates slipped into the village, bringing home sons and daughters. Polystelios emerged from his house and tried to call his flock down from the hills, as he always did, but today they ignored him. Good Friday was not a day to be a lamb in Greece. He went back inside and closed the door behind him.

Come evening, clumps of men gathered in the dusk outside the *kafeneion*, talking in low voices, twirling their *kombolói* beads. Yióryio had offered to take us to the service, telling us to be ready no later than eight, but when we called up to his house he still wasn't dressed. We idled towards the church and Ulysses fell into step with us, hitching up his trousers as he walked, releasing his lunatic babble.

'*Pa papa ma ma boo oo.*'

We spoke to him and he seemed to warm to our voices, encouraging us forward with the beat of his gibbering mantra. He was as excited as a child, in new braces and a clean shirt, subsumed in a ritual which gave him a sense of place, of being an equal in the village.

The white church seemed to glow in the dusk as if lit by heavenly light, aided by the fluorescent tubes which Kóstas had wired up on Thursday. Inside it Christ's bier, the *epitáphios*, was splendid in wild flowers, rosemary and cellophane-wrapped carnations bought in Vrysses. *Yiayiáthes* in black scarves, black aprons and black stockings settled in the choir stalls which lined the body of the tiny church. Behind them a reproduction *Anastasis* icon showed Christ in hell, dragging Adam and Eve out of their tombs like Theseus rescuing Persephone. Ulysses, face shining, pointed at the black-draped altar and then, like a young king claiming his rightful throne, took his seat between the crones.

'When does the service start?' we asked men waiting outside in the shadows.

'Nine o'clock.'

Yióryio opened the *kafeneion* for an hour. We sat at a table in the corner and watched the family reunions. Two elegant Athenians stopped at the bar, tall and vain, the first women to drink there during our stay apart from Katrin and Ariadne. Their husbands stood with fathers and uncles on the threshold, looking out over their valley, in silence. A stranger bought us a drink. The church bells mourned the passing minutes. It felt as if Christ were Anissari's own, only-begotten Son.

At nine the men put down their drinks and set off by foot for the church, amiable hands on shoulders and backs. The Athenians followed behind them in a Mercedes with its radio blaring. Sophia wiped down the tables and Leftéri rearranged the chairs while Yióryio sprinted away to feed Rosetta.

As we waited for him Ariadne drove into the *platía*, returning to her lost home, unexpected and welcome.

'We walk together,' she said to us. Apostoli wasn't with her.

Messa Anissari gathered in the stony churchyard, its men and women in sombre colours, spilling through the open doors of the church. Boys heaved on the bell rope, crashing out a pealing din, and red-caped girls milled between relatives. Iánnis was reeling. Kóstas rewired a fused light. Polystelios clattered up in his *mechaní* and heaved Aphrodite into her walking frame. One of the Athenians answered her mobile phone.

I told Ariadne that we hadn't expected to see her.

'Let's say that I was never the best in religion,' she said as we walked together towards the metallic chanting. 'I was pushed and beaten so much to go to church that I couldn't become a religious person.'

But she had come.

Inside the elders chanted a *mirológhion*, or traditional dirge, passing a microphone from men to women, back and forth, clearing phlegmy throats between amplified verses.

'*Kyrie eleïson*,' droned the old men in a rising invocation, identifying themselves with the figures of their own Eleusinian Mystery. '*Kyrie eleïson*.' Lord have mercy.

Their wives sang a shrill lament for lost sons, casting themselves as Mary in the Easter drama.

'My most beloved son, my sweet springtime, to where has your beauty fled?' they wept for Christ in words that may have been drawn from the ancients.

'You need candles,' said Yióryio, puffing up behind us in his short-arm penguin walk, determined to remain our host. He had dressed in a musty leather jacket and aftershave. He pulled us away from Ariadne, pushing Sophia aside as she shuffled into the church to collect half a dozen slender tapers. 'Ariadne not believe in Big Boss upstairs,' he warned us before explaining that the brown wax candles were more powerful than the white.

Nikos, the Winged Priest, adopted his most pious expression, drooping his lips under his moustache, fancying himself as an icon's saint. He waved his silver ceuser and looked resplendent in a gold and green robe which hung stiff and flat from his shoulders. Aphrodite took the microphone and sang a verse but lost her breath and had to be helped to a pew.

Yióryio and Sophia kissed the Bible, lit their candles from those of their neighbours, then passed the flame to us. We twisted our tapers into the sand of a vast, gilded tub of waxen light and I said a private prayer. Around the bier, around the priest, circled the parish, as in every *koinótita* in Crete, re-enacting survivals of the cult of Adonis. The villagers paused to talk to friends, to kiss, to shake hands as warm brown wax ran over their clasping fingers. They circled too around Ulysses, his eyes filmed over in rapture, seated between the men and women, the gods and Christ, the touched man safely held at the centre of the rite.

Then with the same gesture he used to spread vinegar on salad, the Winged Priest began to spray holy water over the congregation. Yióryio, Socrates, bitter Iánnis and Manólis the carpenter heaved the carnation-covered bier onto their shoulders, swaying as Iánnis found his footing. The village girls, as myrrh-bearing *myrrofóres*, picked up their carrier bags of blossoms to strew over the dead Christ, His holy body represented by a new Bible from Heraklion. Sophia combed Leftéri's hair.

The priest led the *epitáphios* out of the church and the congregation fell in behind him, though not into his sombre piety. Instead of ceasing to chat, they shouted to neighbours, lit cigarettes, even rang hand bells. Their children ran ahead, throwing more flowers at the bier, their voices rising in the anticipation of the promised resurrection. The funeral train wound through the dark lanes, passing every house in the village, casting flickering shadows on their white walls. Nikos' censer clanged and rattled. The smells of incense and jasmine mingled in the night air. In the *mechaní* Aphrodite brought up the rear, her candle held in front of her like a pale beacon.

Only Ariadne kept herself apart. Katrin and I walked with

her, holding a fresh set of brown tapers, slowing our pace as the pious kissed the priest's hand and the superstitious ducked beneath the bier.

'We come together to remember,' she said to us. 'We do everything the way our parents did for their parents.'

'Have you written the names of your dead?' called Aphrodite, leaning out of the cab to grip my arm. 'We remember everyone lost on Souls' Day.'

'I wrote out the names of my lovers,' yelled Polystelios through the racket of the engine, accelerating towards the front of the crowd.

Aphrodite struck out at him and there may have been tears in her eyes.

The convivial, chattering parade emerged from the confines of the narrow streets, moving into the open and towards the smaller, second village chapel beneath three tall cypresses. The priest, bier and congregation climbed the steep steps to the cemetery, pausing for breath as they reached its commanding view of the valley. Nikos processed into the simple building, drooping his lips in lofty devotion and chanting the gospels at the head of the *epitáphios*. Polystelios hoisted his wife onto his back and carried her up over the worn stones. I brought her walking frame.

But the train of mourners did not follow the priest. Instead they broke away and fanned out across the marble and repainted concrete graves.

'Now you see how we live,' Iánnis spat in my ear. 'This is my grandfather.' He pointed at a stone slab over which his mother was sobbing.

'*Theou spitia*,' said Papoos, the high-booted sexagenarian, holding my arm. Two houses. Two graves. 'Mama here. Papa there.'

A hundred points of candlelight brought the cemetery to

life, making it a living part of the village. Tapers burnt on limed stones, lighting carnation wreaths and framed photographs of the dead. An old man took the rope of the cemetery bell and heaved on it with all his heart. The wailing of the women rose like a flood over the graves. Katrin began to cry. In the clamour I heard my mother's telephone number again, ringing over and over in my head, until I could not think.

One new grave was adorned with a snapshot of a young man and a toy motorcycle. 'Remember my friend,' Yióryio shouted in my ear.

As we buried Christ, Anissari's tombs released their dead, rising up as He would the next evening. In his heart Yióryio stood beside his father, as did Sophia, Socrates and Papoos, whose parents had been killed in the bombing of Hania in 1941. Iánnis spilt drunken tears on his mother's grey head. Ariadne squatted alone in a corner by two white graves. The children ate the *kóllyva* from small paper bags with plastic spoons.

In the chapel the Winged Priest read aloud the names of the beloved dead and unleashed another flood of tears. Aphrodite sobbed with such severity that I thought she was having a fit. She fell out of her frame onto a concrete slab and the sound of her keening swelled until it was all but indistinguishable from the hammering bells. The men kept their distance from her, lost in their own grieving. Even Polystelios did not touch her. Instead Sophia waved Leftéri towards the crumpled hunchback and Aphrodite wrapped him in her pallid, fleshy arms.

As abruptly as it had begun the atmosphere changed again. A hush descended on the village. The bearers hoisted the bier in silence and waded through the low mist of tears and candle smoke which had settled over the graves. We wound back to the upper church, hearing only each other's footfalls. Outside its door Iánnis stumbled and the Bible fell onto the cobbles. The pallbearers rebalanced their load then lifted it high above their heads. Every villager walked beneath the bier and through the church, tearing off a flower as they passed, then crept away along the dark streets.

No one spoke for Christ was dead.

We walked Ariadne to her car, drained of words and emotion. She slid behind the wheel and said, 'All my life I fought for a new world; now I cry for yesterday.'

Then she drove away.

At home I didn't turn on the light. We undressed by moonlight, slipping into the hard bed without words. No nocturnal voices rose over the basin of olive trees and, as everyone was indoors, no dogs barked along the road. Only the croaking of tree frogs reached our ears and, from behind the house, the occasional rustling of birds in the scrub.

Around two o'clock I woke to the sound of footsteps and went to the window. It was Yióryio. When he couldn't sleep, which was most nights around the full moon, he'd leave his home and the *kafeneion* 'in care of Big Boss' and tramp around the village. He'd pause outside houses, thinking of the residents inside, making a kind of blessing: 'Here living Ulysses' or 'Here being new couple where one day may be baby.' It was an odd, paternalistic caring, maybe misplaced or arrogant but nonetheless sincere and better than drinking beer in front of German satellite television reruns. After he passed for the second time I fell back to sleep.

But it wasn't Yióryio who hammered on the door at four o'clock, or three minutes after according to the chime of the Pano Anissari clock. It was Polystelios.

'My wife must go to the clinic now,' he pleaded.

We dressed and collected the car in less than five minutes,

reversing up to their door in case Aphrodite needed to be carried out of the house. She had looked poorly all evening, her skin ashen, and as we hurried we feared the worst. But when we pushed into their front room, ready to apply a tourniquet or to mop her brow, she sat composed and upright in her chair, wearing her best skirt and jacket. She had even brushed her hair.

'We must go now,' said Polystelios, though Aphrodite seemed in no hurry, taking a moment to look around her home and offering us coffee.

Together we carried her to the passenger door but she insisted on sitting in the back with Katrin, holding her hand during the journey. Polystelios perched beside me on the front seat, his nose pressed against the windshield, shouting out unnecessary directions at every hairpin bend on the Vamos road. Ours seemed to be the only sounds in all Apokoronas; the rattle of the diesel reflected off low stone walls, Aphrodite's small-bird intimacies whispered to Katrin, Polystelios' anxious bellow up to the doctor's house which woke half of the town. The doctor came out to the car, flashed a torch in Aphrodite's face then told me to drive on, taking Polystelios' place beside me.

At the clinic he led us towards his examination room, switching on all the lights including a new 'Christ is risen' sign above the town square. The mayor had not wanted it illuminated until after midnight mass on Easter Sunday. Katrin held back at the swinging doors but Aphrodite clasped her hand and pulled her inside, along with a nurse who appeared from out of the night. Polystelios gestured for me to stay outside with him.

'It's her heart,' he said to me, pacing back and forth across the square, looking more unwell than Aphrodite under the strobing, multi-coloured neon. 'She has a weak heart.'

Her heart, which was soft rather than weak, wasn't the problem, but gallstones that had brought on obstructive jaundice. Her porky frame and sedentary lifestyle meant that the symptoms had gone unnoticed for months, or had been linked to other ailments. Her yellowing skin had been attributed to their appalling wine and domestic hygiene. In throwing herself about the cemetery Aphrodite had dislodged the stones, which had then compacted in the bile ducts. That was the doctor's diagnosis, even though he could feel nothing with her gallbladder so inflamed.

The Vamos X-ray machine was broken and the operator was in Lassíthi for the weekend. Aphrodite had to be transferred to Hania, where the technicians were doubtless also on holiday, because surgery was essential. The doctor tried to call for an ambulance but the driver had turned off his mobile phone. Who would dare fall ill with Christ already dead? We all squeezed into my car, apart from the nurse who went home to bed.

On the deserted National Road I nearly ran over a skipping lamb which had survived Easter's butchery. At the cream-and-blood-red Hania hospital the surgeon announced that the gallbladder had to be removed, which seemed such an extreme diagnosis that we put it down to mistranslation. Katrin dug out our learners' dictionary but there was no misunderstanding.

'Cut it out,' he said with chilling clarity. 'It is stale.'

Aphrodite moaned, wept and enjoyed being at the centre of attention, even if it meant losing body parts.

But the operation couldn't be performed until Sunday. Polystelios paced back and forth across the examination room, torn between staying with his wife and attending to his chickens and sheep. He scoffed at my suggestion that Yióryio could look after the animals. When he was offered a spare bed on the ward, vacated by a cancer patient who'd gone hunting for the weekend, he decided to stay at the hospital.

'How much do I owe you?' he asked me.

In front of the doctors he made a grand gesture of reaching for his lean wallet and offering to pay for the lift. I refused and he looked both relieved and offended. As poor as he was he could not be seen accepting charity from a foreign guest. *Filoxenía*, the honoured 'love of strangers', was not simply a matter of Homeric hospitality, but also a demonstration of the host's superiority. Polystelios began thrusting notes into my hand, offering me the money he'd earned selling his lambs. When I declined the payment he dug the coins out of his pocket.

I put my hand on his arm. 'Your wine is worth more to me than money,' I lied. 'One bottle of it will be payment enough.'

'You will have six,' he declared, preserving both his savings and *timí*. Honour.

At least I'd be able to unblock our drains.

We drove east into the Easter dawn, dropping the doctor off in Vamos. Christ was dead, Aphrodite sick and in the villages shards of crockery were thrown out of the window to harass Judas.

The shattering of death, or at least of the past year's plates, had changed Anissari. The men were in the fields cutting dead branches to prepare a thirty-foot-high bonfire in the churchyard. The women were indoors again, making *mayirítsa*, a soup of tripe, rice and lemons, to break the Lenten fast. The bread van delivered Easter cakes. Children filled the church with laurel and the muffled excitement of anticipation. I stayed out of the garage, as Yióryio had insisted, helping him instead to muck out Polystelios' animals.

In the evening we picked our way over the broken crockery back to the church, together once more with all the villagers, except Aphrodite and Polystelios. I had made him promise to ring me so I could collect him from the hospital. I hadn't considered that for him to do so would have meant losing face.

The Easter Sunday mass was the high point of the Greek calendar. Tradition called for the priest to douse all the lights at midnight, plunging the village into darkness and symbolizing 'Christ's journey through the underworld', morphing legend into biblical parable. Then, out of the night, the priest would appear bearing a single candle. He would pass the flame to his parishioners, taper touching taper, the consecrated light rekindling faith in life, again and again, until all within the dark body of the church were bathed in God's light.

'This is the light of the world,' the priest would chant. 'Come take the light.'

However, the Winged Priest was in a hurry. He had three other midnight masses to conduct so he left the electricity switched on and some time after eleven o'clock simply walked to the door with his candle.

'*Afto to fos*,' he muttered, without much feeling and trying not to look at his watch. '*Théphte, lávete fos*.'

'*Christós anésti!*' shouted the crowd in response, wasting no time for they were impatient too. The pyre was already alight. The flames leapt up through the broom and the wood. Sparks swirled into the night sky, singeing the electricity cables. The petrol-soaked effigy of Judas exploded into a plume of flaming straw and the burning torso landed on the co-operative's transformer which fused half the village. Aerosol cans popped like gunfire, releasing a fragrant steam of deodorant and hair spray.

The scorching heat drove us back towards the bells, which

rang with fantastic enthusiasm. '*Chrónia pollá*,' shouted friend to neighbour, brother to aunt. Many years. Long life.

'*Epísis*.' And the same to you.

Yióryio pulled a nine-millimetre pistol from his jacket and fired above our heads. Rifle rounds echoed off the houses in the *platía*. A crackle of fireworks added to the racket.

I caught the Winged Priest as he strode to his car and thanked him for letting us attend the mass. 'It's a lot of work,' he grumbled. 'I have to do four services.'

As the bonfire died, villagers walked home, cupping their candles to make a sooty black cross on the lintels above their front doors. Katrin and I fell into the *kafeneion*, which had been closed to customers but opened for its extended family: loony Ulysses, his superstitious mother Chryssoula, the widower Papoos, one-armed Iánnis and ourselves. Sophia arranged the tables end to end and laid a dozen places along the white tablecloth. Leftéri and his sisters helped to set out Easter bread, boiled potatoes and spleen soup.

'In Crete we fight angel of death,' joked Yióryio, tossing a hard-boiled red Easter egg to each guest. 'Of course we not win but he get fucking good hiding.'

We cracked the eggs against one another's, the first champion challenging the next, until only Yióryio's egg survived unbroken, which boded well for him in the year ahead.

'I have luck because I am strong beyond measure,' he laughed at me. There was no whisper of the previous night's emotion in his voice. His mourning had been purged, his soul cleansed, at least for another twelve months. 'We share *mayirítsa* as we share my luck, *pilóte*,' he added, with his usual imperfect logic.

Little Iánnis served himself and ate with his head down, locks of hair hanging in the soup. Chryssoula took a bird's portion of salad. Ulysses chewed his egg without removing its broken shell.

As the food was passed around the table Yióryio rested his hand on my shoulder, 'This evening we live, we eat, we make babies, for no man is dead . . .'

'Not yet,' said Iánnis, lifting his arms as if wings.

Yióryio ladled yellowed spleen into my bowl.

'. . . and before anyone dies you will gain ten kilos.'

'Then be too fat to fly your aeroplane,' said Papoos.

In Crete they kill you with hospitality. After the late supper we awoke well filled and decided to walk off our heaviness. On the map we traced a long ramble from Anissari to Gavalohori, with its mulberry trees planted by the Turks, and down the valley to Almirida's sandy, protected bay. We gathered together our boots, drinking water, two peaches and set off by ten o'clock. Twelve hours later we hadn't even reached the edge of the village.

The morning air was laced with smoke, both from the Easter pyre and a dozen cooking fires. Yióryio caught sight of us first and called us into his vegetable garden. He had filled a pit with charcoal and suspended a whole lamb above it. We took turns at winding the spit, at a clip because of its closeness to the embers, and drank coffee together. Sophia brought out slabs of cake to sustain us in our labours.

An hour or so later we extracted ourselves, promising to stop by on our way home. We walked twenty yards along the road and were hailed by Kóstas. With his bulldozer the policeman had gouged out a trench in a vacant lot to cook two sheep, as well as a copper-coloured sausage of offal like a metre-long kebab. His brothers and their wives were home for the holiday. They all insisted that, as guests of the village, we have the first taste. We tried to refuse them but relented after a glass of his excellent wine. Then we drank another as the lamb wasn't quite ready. At noon Kóstas cut two thick slices of burnished flesh. It was so delicious that we each accepted a second piece, washed down by a third – or was it fourth? – glass of wine.

Sometime after one o'clock, though it could have been later, we set off again with every intention of walking to the sea, or at least to a shady tree in Gavalohori. But another plume of aromatic smoke rose up into the clear spring sky. Manólis the carpenter had dragged an old metal bed frame from his workshop and raked underneath it embers from the bonfire to create a makeshift roasting platform. He and his family – which seemed to include all of Exopoli's geriatrics – invited us to sit with them. They too offered us more than we could eat, as did half a dozen other villagers who called us to their doorways and coaxed us to join in their meals.

Our afternoon passed in a daze of eating and drinking. We could refuse no offer of food and were bound to confess aloud that each mouthful of lamb was the finest we had ever tasted. At house after house we accepted yet another *small* plate and glass of wine. I tried to bring some purpose to the day, other than gluttony, and asked our various hosts about engines. Socrates the shepherd offered me his old mini-tiller, with its two-horsepower two-stroke. Manólis' brother from Sfakia had a 5½-horsepower Honda generator with electronic ignition. Papoos remembered, his voice lifting with excitement, that his cousin had a new chain saw. Their offers were spontaneous, and given time I might have been able to adapt them to the Woodhopper, but any conversion would have made them unsuited to their original purpose. So I resigned myself to celebrating Christ's victory over death and ate like a pig.

Around six o'clock Polystelios' return to Anissari snapped us out of our sated reverie. He limped into the fields like the grim reaper, his scythe balanced on his shoulder. Thirty minutes later he returned under the weight of a bundle of freshly cut fodder, walking with his sharp, arthritic waddle. We learnt that Aphrodite was to undergo surgery on Tuesday and Polystelios, concerned for his animals, had tried to catch a bus home. But there was no public transport on Great Sunday. So, rather than call me, he had hired a taxi and emptied his wallet on the ride to the village. We stopped in to see him and, even though he had only eggs for his Easter meal, he refused to join his neighbours' feasts. He wanted nothing and his neighbours let him be.

In the evening lyre music wafted across the tops of the olives. Shepherds and beekeepers, *yiayiáthes* and young boys danced in the *platía*. The villagers' celebration of Easter, their mourning and rejoicing, taught me that Cretans, possibly more than any other Europeans, were rooted to their land. They were physical people of light and weather, bonded to mountains and sea. This raw affinity freed them from pretence. The first time I'd used this word Yióryio couldn't grasp its meaning. 'But why anyone pretend to be another?' he asked me. 'Everyone would know him – his mother, his wife, his friends – and know what man he is.'

I walked into the fields. In years past at home in Canada and Britain, Easter had never moved me. The story of death and rebirth had lacked poignancy. But now, walking alone through the vines, chased by distant music and personal grief, I felt shaken and disturbed. In the midst of the celebration of life eternal, I began to accept the finality of death and its place in life. After bereavement the world never again is whole but in a shared outpouring of grief, in collective mourning, in the spring flowers blooming over a lush Cretan valley, the dead could be reborn in one's heart.

I took the back lane to the house, trying to keep out of sight, but Yióryio spotted me and called out.

'*Yá sou, pilóte*,' he shouted, wide of shoulders and face. '*Poú pígate?*' Where have you been?

There was still roast lamb to eat and, didn't I know, his was the very best in all of Crete.

18. Lease on Life

After Easter I felt as if I'd been granted a new lease on life. That first Tuesday was like the start of a new year. The morning seemed to dawn earlier, brighter, beneath a warming sky and wisps of cloud which laid soft, slipping shadows across far green hills.

I was up at first light when the air was still fresh and the earth pearled with dew. My morning walk stretched a line of footfalls across the grass. Black bees fed on red poppies. Yióryio's guinea fowl pecked through the garden, gurgling under the figs. Finches and larks called from the eaves, darting into the olives which twisted and sighed in the morning's breeze, turning silver-soled leaves towards the sun. It would be hot in an hour, but in those first cool, calm moments farmers idled out to their vines while their wives collected snails along the roadside. Their sons and sons' wives shook themselves awake, drank their first coffee on their feet and drove to Vrysses and Hania, to office desks and tour agencies.

There were still major obstacles for me to overcome but with the sun's warmth came a new optimism, as well as a measure of Cretan acceptance. If my problems weren't solved today then they would be tomorrow, or 'after tomorrow', for, as Apostoli assured me, 'the God has it'.

In the garage I sped ahead with the wiring, rigging the tail assembly and balancing rudder to elevator. It was meticulous work, precutting six wires, adjusting their tension then swaging – or clamping – the small, copper Nico sleeve anchors. But the morning's labours brought me an unexpected sense of satisfaction. After all, it was a rare privilege to pursue a childhood dream into adulthood. I was building my own aeroplane and, whether the machine flew or not, I found myself enjoying its creation. At last.

All of a sudden there didn't seem to be enough hours in the day, or lives in the body. As the aircraft took shape I saw that I was taking leave of my mother. I knew that I might end up smeared along the runway, and, if that was to be the case, so be it. But I believed that my end would probably be much more ordinary than that and the time had come now to make something in my life. I would not leave behind an ample inheritance or a useful invention but I could draw a fanciful line across the sky with a fragile, hand-made flying machine. And that line underscored my two reasons for living: first, to love and be loved; second, to mark, record and remember that love.

I recalled the story of a friend who had also lost her mother. Night after night she had tried to reach out of her sleep and mourning to ask her, 'Why did you leave me?' Finally, after months of calling her name and hearing no response, the mother's phantom appeared before her in a dream and said, 'Will you *please* leave me alone. I'm dead.'

★

Little Leftéri had a new model Junkers 88 which he flew around the garage all morning, dive-bombing my thoughts. I chatted with him and around twelve o'clock stopped to share an early lunch, both of us tearing fist-sized chunks off a loaf bought from the bread van. Sophia disapproved of our meagre meal and brought us each an omelette.

'Do you want to be a test pilot when you grow up?' I asked him.

He looked at me askance and said, 'I want to be a bulldozer driver.'

In the afternoon I stopped at the *kafeneion* to pay the second month's rent. Socrates the shepherd counted the notes twice then slipped them into his breast pocket. Again he promised to give me a receipt but I understood now that none would be forthcoming.

'You mentioned a mini-tiller,' I reminded him.

'My brother-in-law took it to his village,' he said. 'He will return it next week.'

Which meant that the loan was a non-starter.

'How about the chain saw?' I asked Papoos, his nose buried in his cards.

'You can borrow it any time,' he said, sober now and not looking up from the game, 'after I've cut the olives.'

Which meant not before the Nativity of St John the Baptist in late June, at the earliest.

'My cousin sell you Fiat motor,' said Yióryio, rifling through the open till by the telephone. The card players' empty beer bottles were lined behind him like a kind of Heineken abacus. 'It make twenty-three horsepower and can't go wrong,' he read from a scrap of paper. 'Except need new fan.'

'Do I need a fan?' I asked.

'Maybe in summer,' suggested Kóstas, who seemed to spend more time in the *kafeneion* than out on duty.

'But the propeller will keep it cool,' I said.

'Then you need only starter motor,' announced Yióryio, 'which you can take off old car.'

'I couldn't start it by hand?'

'A Fiat?' laughed Socrates, who knew about such things. '*Poté ton potón*.' Never of nevers.

'Also there is no fuel tank,' added Yióryio, looking down at his notes.

I wanted as simple and as complete a kit as possible. An automobile engine would probably be both too complicated and too heavy. But if I stripped off the outer cowling, jump-started it with a car battery and kept an eye on the temperature it might work.

'Where does your cousin live?' I asked Yióryio.

'In Athens.'

An overnight ferry ride away.

'What does he use the engine for now?' I asked. Without a fan, starter and fuel tank.

'Nothing,' Yióryio replied. 'It broken. Why else for sale?'

After three months in Greece I'd learnt to keep my options open. In the STN Atlas executive washroom I'd found a back issue of *Flight* magazine and drawn up a list of aeronautical manufacturers in Europe and America. I bought a handful of phone cards over the bar, set a chair in the sun by the public telephone and started dialling.

I called GKN Westland first, not because the Woodhopper was anything like a 2,800-horsepower Apache attack helicopter, but because their headquarters were near my home in England. I hoped they might be willing to support a local project, even one relocated to Crete.

'Our engines are of the gas turbine variety,' said Richard Case, the chief executive, after I'd outlined my needs, 'so rather too big for your flying machine.'

Case had nothing smaller collecting dust in a forgotten corner of the site and advised me to approach Meggitt Aerospace in Hampshire.

'Don't fly too close to the sun,' he added.

The Greek operator found me the number, having first put me through to Maggot Pest Control in Hampstead, but Richard Grieves, the managing director, was away in Switzerland. I left him a long message, interleaved with the crowing of Yióryio's rooster, and went back to my list.

Thruster Air Services in Wantage weren't interested in helping me. Sky Systems of Brighton suggested a go-cart engine. Paul Owen, an amateur aircraft builder in Dorset, thought he might have something handy but was too busy to go down to his shed to look.

'Try to get a hold of Eric Clutton's book *Propeller Making for the Amateur*,' he advised me.

'To carve my own prop?' I asked, never having considered the possibility.

'Why not? Eric would have done it for you but he's moved to the States. He was catching a cold with regulations here.'

Paul passed me on to the Aircraft Restoration Company at Duxford airfield.

'What's your prop clearance?' asked Jim Romain.

'I don't know.'

'How can you not know?'

'I haven't built that bit yet.'

Jim didn't want to do business with me. He seemed to think I was a nutter.

Next I called the British Microlight Aircraft Association.

'Are you the fellow trying to kill himself in Greece?' asked the operator. It seemed that news was spreading. 'I suggest that you make the inaugural flight *after* writing the book.'

As I made my calls Yióryio brought me glasses of wine. I

might not have been doing well but I was having a good time. I bought more phone cards as Papoos flagged down a passing cheese vendor. He cut a cake of *mizíthra* into door-stopper slices. My chunk of cheese was the size of a half-pound pat of butter.

Through the afternoon I called Lindhurst Touchdown Services and Verdict Aerospace. I spoke to aircraft importers about Quicksilvers and Simonini Tango 2s. In Warwickshire I found an Italian pusher engine but there was concern about reversing the bearings. So I rang the manufacturer in Milan.

'Will it work as a tractor?' I asked the sales manager.

'You want to put our engine in a tractor?' he shouted, insulted by my question. 'We do not make agricultural machinery.'

In Cornwall I found the owner of a defunct Second World War air base, his hangars cluttered with old junk. He thought he might have an rusty Hero engine and agreed to hunt around the back of the leaking sheds.

'I'd be lying if I told you it's any good,' he reported when I called back. 'The hangar roof caved in fifteen years ago. Could you make do with a boat outboard motor?'

I cast my net wider, calling a French home-build owners' club. Anne Vandamme worked for Alcatel Space Industries in Cannes during the week and at the weekend was building a monoplane called the Souricette.

'Of course I know the Woodhopper and la Demoiselle,' she enthused from her home in Fréjus. 'These are planes I love.'

But she knew no one who would loan me an engine.

'Even a used one?' I asked.

'Not new, not used, for perhaps it come back in very small pieces, *non*? And do not forget you will need a reduction system too.'

'A what?'

'Perhaps my husband and I should to come to Crete to help you?' she suggested. 'You know I am qualified to teach aeronautical lessons to children.'

Anne offered me the number of Joe Cauthen, one of the Americans who had updated the Demoiselle's design. I caught up with him eating his breakfast in Clinton, Missouri.

'I cannot believe the Woodhopper's still alive,' he said.

'Both it and me,' I assured him.

I reminded him that his original plans had called for a Chotia 460-C, an engine which no one knew any more.

'Nothing else was available to us during that earlier period of time,' said Joe. 'Today you should try a Hirth 310. It'll run a long stretch between overhauls.'

In *Flight* I found the number of a Hirth distributor in Germany. I called Hanover and spoke to Helmut.

'You are building an aircraft in Crete?' he asked, after I'd explained the situation. 'With Greeks?'

'Yes.'

'And you are surprised to be having trouble?'

'Many people are helping me here,' I said, hoping to allay his misgivings. 'Trying to help me.'

'You know the Greeks do not have a sound reputation as aircraft builders. They are more put-the-lamb-on-the-stick-and-get-very-drunk type of guys.'

Helmut wouldn't loan me an engine.

Yióryio brought me yet another glass of wine. I told him what I'd achieved, which wasn't much.

'*Then peirazi*,' he said. It doesn't matter.

The rest of the day slid away in companionable simplicity; talking, eating grilled chunks of lamb's liver, dozing in the heat of the afternoon until the men went off to plant vegetables and light the lamps.

Around seven o'clock – in an amiable frame of mind – I decided to have one last try and called the number of an organization called UAV in Virginia.

'NASA Wallops Flight Facility,' answered the operator.

'NASA?' I said. 'The space agency?'

'That's right.'

I checked my notes. 'I'm after something called UAV.'

'One moment please.'

UAV turned out to be an acronym for Unmanned Aerial Vehicle. The Wallops base was in Chesapeake Bay, just up the coast from Kitty Hawk, North Carolina, where the Wright brothers had first flown. The NASA facility there, with an interest in both surveillance and military drones, acted as a kind of Consumers' Association for the aviation industry.

'What's your mission profile?' asked the captain who answered the phone.

'Mission profile?'

'What's your intended range, altitude and flight duration?'

'Not far,' I said. 'And not too high.'

To keep him on the line I told him that I was a Greek zoologist who needed a surveillance drone to study mountain goats.

'Your best bet then would be a Theseus or Perseus,' suggested the captain with sincerity.

It pleased me to hear that Greek gods flew for the American aerospace industry.

'Theseus operates on either a twenty-eight- or a fifty-kilowatt power plant.'

I didn't know anything about kilowatts.

'The smaller one sounds safer,' I told him. And easier to control. 'What does it cost?'

'The short-life version runs for forty hours and costs $3,000.

Double that for the long life. And they'll throw in the propeller.'

I hung up the phone. I was out of phone cards and my ear hurt. It was time to go home.

I'd walked no more than twenty metres when the telephone started to ring. I sprinted back, grabbing the receiver from Ulysses' hand before he started to babble into it.

'Hello?'

'Mr MacLean? Stuart Westlake-Toms of Meggitt Defence. We're interested in your project.'

According to *Flight* Meggitt Defence was a world leader in the design and manufacture of free-flight aerial targets and unmanned aircraft. Its parent company, Meggitt Aerospace, made vibration monitors for Boeing, engine sensors for Rolls-Royce and an 'Auto-Fault Digital Fuzz-Burn' system for the US Coast Guard's helicopter fleet. My voice-mail message had been sent from Hampshire to the Swiss HQ and back to the shop floor in Kent.

'We're prepared to assist you with the loan of a new engine,' said Westlake-Toms, the business development director. 'But we insist you sign a disclaimer as they're not certified for manned flight.'

I told him I'd sign.

Meggitt's drones operated in thirty-eight countries. They were powered by a range of engines but their two-stroke, twin-cylinder 342-cc unit would suit the Woodhopper, especially as it weighed under twenty pounds.

'We've made thousands of them,' said Westlake-Toms, a man whom I'd begun to like enormously. 'We fly them for two hundred hours before retiring them. They're popular in one-man hovercrafts too.'

'I hope to fly a bit higher than a hovercraft,' I told him.

Westland-Toms suggested that I visit him in Ashford to see

an engine running. But a trip to England only made sense if I could get a propeller there too.

'I left a message with a prop maker called . . .' I hesitated, scrabbling through my notes, '. . . Dennis Nixon.'

'Dennis is a former owner of Meggitt,' said Westlake-Toms. 'Flying is his passion. Shall we say he can't afford a new Bentley because of the upkeep on his Mustang, plus his six other aircraft.'

Westlake-Toms thought Nixon would have a suitable propeller for me.

'I'm off to Kuwait at the end of the month,' he said. 'Why don't you fly over and collect an engine and prop before then?' I reached for a chair. I had to sit down. 'I'm sure you'll enjoy the visit. We're something between a high-tech aerospace company and Morgan cars.'

I put down the phone and, in a state of shock, wandered into the *kafeneion* to join Yióryio. I told him about Meggitt.

'Nothing ever falls on the ground,' he said, refilling my glass. 'I giving you luck.'

An hour or so later I started for home, again.

In the grass outside the *kafeneion* I spotted the perfect pair of wheels, attached to an ancient, rusted moped.

'Who owns the *mechanáki*?' I asked Yióryio, now leaning against the door and gazing out at the mountains.

'Spiros.'

Spiros was inside watching television. 'Can I have it?' I asked him.

'I'll sell it to you.'

'But you must have thrown it away ten years ago,' I said. Or longer, judging from the rust.

'And now you want it so it's for sale.'

'How much do you want?'

'What do you have?'

I opened my wallet. 'I have eight thousand drachmas,' I said. About £15. 'I need three thousand to buy petrol so I'll give you five thousand.'

'Sold,' said Spiros, returning to the television. 'But you take it away yourself.'

It had been a good day.

19. Building Icarus

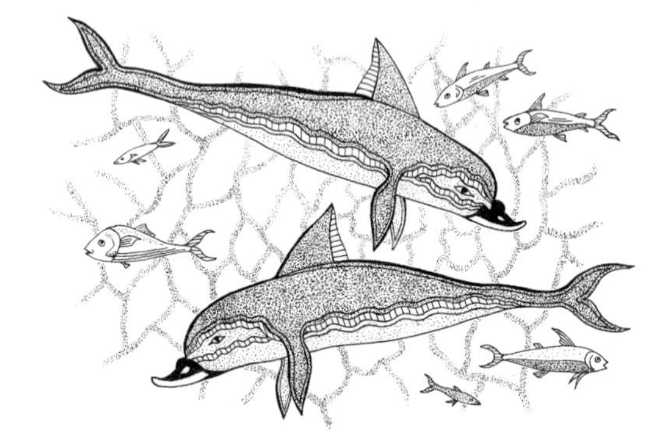

Crete was both Zeus's mythical birthplace and Homer's 'fair, rich and sea-girt land in the midst of the wine dark sea'. It was Lord Howard's 'witches' cauldron' and Lonely Planet's 'paradise of long sandy beaches and isolated coves'. The 'blessed' island was 'the isle of unrest', reinventing itself through the ages. It was a place of dynamic, expedient creativity, not so much in the work of its contemporary artists, but in its beliefs, in the fabric of its society, even in the bedrock of the island itself.

The Minoan gods had been co-opted into the Greek myths, renamed by the Romans, then re-emerged, to begin again, as part of Orthodoxy. The Earth Mother became Hera, Athena then was swallowed up in Mary, mother of Christ. Perhaps the twelve Olympian gods were conjured into the twelve apostles by a sleight of hand.

'How can they be cousins?' teased Papoos, winking as – like most historians – he denied the seductive continuity. 'Zeus was Greek and Jesus was a Jew.'

Cretans became Greeks, shifted their allegiance from Rome to Constantinople, drove out the Venetians by appealing to the Turks, who abolished forced labour as a quarter of the population converted to Islam, changing yet remaining unchanged. Two centuries later Venizélos exiled the occupiers, as Muslims reconverted to Christianity or sold their property for the equivalent of the boat fare to Anatolia. Shepherds became royalists then Communists, traded futures on the Athens Stock Exchange and, on the profits, their sons built hotels to welcome back the foreigners.

Beneath the shallow, concrete foundations of the beachside resorts, the tectonic plates collided, transforming the land above. Knossos was levelled twice in ancient times by earthquakes and tidal waves, only to be reborn as Europe's most visited archaeological site. Heraklion's imposing St Francis monastery, destroyed in the 1856 earthquake, re-emerged as the Archaeological Museum. On the east coast the Minoan palace of Zakros dipped another inch under the waves every decade, or eight metres over the last two thousand years, while on the western tip of the island a sublime, curved beach rose out of the sea.

The first decade of the twentieth century was also a time of invention and reinvention. Queen Victoria died, Prince George sailed away from Hania and the Cretans argued towards union with Greece. In 1909 Freud lectured in the United States on psychoanalysis, Blériot made the first cross-Channel flight and Santos-Dumont began selling his monoplane to the public. In the same year the archaeologist Arthur Evans neared completion of his Knossos excavations.

Evans, the son of a wealthy English numismatist, was 'a man of paradoxes; flamboyant and oddly modest, extravagant yet by no means self-indulgent and in some ways austere'. At the age of forty-three he set out to discover the truth behind the Minotaur myth.

It had long been accepted that the Greek myths were, in the words of historian George Grote in his twelve-volume *History of Greece*, 'essentially a legend and nothing more'. Greek history began with the first Olympic Games in 776 BC. As historians had no means of distinguishing fiction from fact, all earlier history – the Trojan Wars, the hero Agamemnon, the *Iliad* and the *Odyssey* – were consigned to the realm of myth. Until 1871, when Heinrich Schliemann, the German amateur archaeologist, found ancient Troy, proving that tangible truths underlay the Heroic age.

Like Schliemann, Evans believed that the Western world's oldest stories were based on real people and places. He too wanted to push back the bounds of history. He bought the suspected location of ancient Knossos, started digging in the dirt and found the remains of Europe's first city.

Few people discover new civilizations; fewer still then reinvent them. Evans was determined not only to reveal the remains of Minos' palace but to 'reconstitute' it. Across the 5½-acre site his audacious restorations, while preserving the Minoans' unique architecture, reflected his imaginative theories. He replaced tapered columns with Art Deco-influenced reconstructions, even though no wooden originals were ever found. He gave the palace a fanciful upper storey, the *piano nobile*. He rebuilt the Central Court to recreate its 'spirit' of splendour. From painted plaster fragments he reconstructed amazing frescos of dolphins, birds and 'a lady in a gay jacket with a very good profile' whom he imagined swirling in a dance.

'It is not difficult to believe,' Evans recorded in his journal, 'that figures such as this, surviving on the Palace walls even in their ruined state, may lie at the root of the Homeric passage describing the most famous of the works of Daedalus at Knossos – the Dancing Floor of Ariadne.'

Evans revived a fantastic world then made it real. He created the 'Prince of the Lilies' from disparate limbs and a piece of plaster showing the back and ear of a man wearing a crown. The absence of defensive walls led him to assert that Knossos had been a powerful and peaceful 'thalassocracy' which dominated the eastern Mediterranean throughout the Aegean Bronze Age.

He had his critics, some of whose reputations he tried to destroy, and his wilder declarations were disproved in the end. Evelyn Waugh wrote of the restorations, 'One cannot well judge the merits of Minoan painting, since only a few square inches of the vast area exposed to our consideration are earlier than the last twenty years.' Evans' theory of a peaceful *Pax Minoica* was then undermined by the identification of networks of coastal watchtowers and later by the discovery of the bones of slaughtered children, suggesting the practice of human sacrifice.

But his energy, determination and personal fortune ensured that, according to his successor John Pendlebury, 'without restoration the Palace would be a meaningless heap of ruins'.

'We know now that the old traditions are true,' Evans declared near the end of his life. 'It is true that on the old Palace site what we see are only the ruins of ruins, but the whole is still inspired with Minos' spirit of order and organization and the free and natural art of the great architect Daedalus.'

Like Evans, and in this way only, I felt that I was building on the past, reinterpreting it, creating something new out of old stories. I didn't believe that Minos' wife had mated with a bull and produced a beast which Minos, in his shame, had locked away in the labyrinth. For me the labyrinth was an allusion to the circular nature of life with the beast – brutality and death – lurking at its centre.

But my rational interpretation held none of the compulsive power of the myth. We need myths, the distillation of hopes and values, to give sense to that which precedes us and that which will outlast us. We want their enchantment to be true.

Evans would have been a contemporary of Yióryio's unnamed, would-be aviator. I imagined that news of the Knossos discoveries could have reached him in England. Such a man might have read Evans' report in *The Times* that the excavations had brought to light 'the oldest throne in Europe'. He could have heard the archaeologist's claim that the rough limestone statue of Mother Rhea and the infant Zeus was the precedent to the Christian Madonna and child. Maybe he was moved – like many others – to make a donation to Evans' Cretan Exploration Fund.

I fleshed out my private myth, influenced by the villagers' tall tales. I fancied the aviator as an engineer, cartographer or auditor, posted to the island for six months with Sir Edward Law's Cretan Commission. In Hania he could have been put up by the British Consul Esmé – later Lord – Howard, who rode out with him to meet local landowners. Together Howard and the aviator might have planned the siting of a new bridge or valued an olive wood which had been cut down by Turkish troops during the fighting, then submitted the landowners' claims to the Indemnity Fund. Their Christian hosts would have offered what they could afford, a bunch of grapes or a glass of wine, and ignored the Muslim grooms. They'd sit together in the sun talking politics, Howard translating, as the aviator's modern Greek would still have been poor.

In those days travellers came to know the island on horseback, through fierce winds and hard handshakes, armed insurgents and winter nights spent in front of log fires, the walls and rafters of the village houses glowing with ruddy warmth. They carried with them John Murray's *Handbook for Travellers in Greece* with advice, from Edward Lear who had himself visited Crete, on travelling with few encumbrances:

> . . . a certain supply of cooking utensils, tin plates, knives and forks, a basin, &c . . . two or three books; some rice, curry powder, and cayenne . . . as little dress as possible, though you must have two sets of outer clothing – one for visiting consuls, pashas, and dignitaries, the other for rough, everyday work; some quinine made into pills (rather leave all behind than this) . . .

At the end of his posting I imagined the aviator helping to draft the final report for the International Committee. It took the view that Crete was 'an incessant source of anxiety, preoccupation and useless expense' and advised disengagement. He, on the other hand, decided to stay on.

Rather than return to the Colonial Office or Hydrographic Service, I had the aviator request a leave of absence. There was no one waiting for him at home and he felt at ease in Hania, enjoying the intense physicality of the outdoor life, of men standing shoulder to shoulder, of history being revived along the coast at Knossos. When Ronald Graham relieved Howard as Consul, I moved the aviator into a pair of rooms above the Café de la Port. He engaged a Maltese cook, one of the large colony on the island, and tried to extend his friendships beyond the tight circle of the International *stationnaire*.

He would have played tennis with British naval officers and swum from the Italian Consul's bathing hut at Souda Bay. He could have witnessed the arrival of the Russian fleet, under Admiral Nebogatov, when it called at the island to revictual. Several hundred crewmen were given leave to visit Hania and used the opportunity to drink the town dry. Their wretched bodies were strewn along both sides of the Souda road. Every wheeled vehicle in town was mustered to haul away the miserable men, all of them peasant farmers who had never before been to sea. They were thrown one on top of the other like so much baggage and witnesses watched half of them drown when two launches taking them to their ships capsized.

The joy following the evacuation of the Turkish troops in 1898 had evaporated and Cretans were now anxious to see the end of all foreign intervention in their island. 'Murders and other outrages perpetrated by some bands of Greek insurgents belonging to a rude highland race' were reported in the *Illustrated London News*. The insurgents had developed a taste for seizing small government custom-houses, for little gain other than fleeting notoriety. The aviator could have accompanied Captain Tupper of HMS *Venus* to recover one building which had fallen into the hands of an insurgent named Birakis. The afternoon was too hot for a fight so, after declaring his determination for Hellenic union, Birakis handed over both the custom-house and all the money in the till, which was not usually the case.

The more common outcome, at least in the Russian zone, was bloody. When the Khrabry *stationnaire* was sent on a similar mission and its men were fired upon, the Russian captain ordered the bombardment of the village, causing extensive loss of life.

In the French sector officers dispatched to reclaim official buildings tended to be found later dining with the insurgent leaders. The Italians on the other hand preferred not to take any action at all. Baron Fasciotti, the Italian Consul, favoured 'a policy of masterly inactivity as far as internal Cretan dissensions were concerned', according to Lord Howard.

The aviator would not have needed to work. Once converted into Cretan drachmas a handful of pounds would keep a thrifty man for a year, maybe two. He could spend his days pottering among the ruins of Kastelos and Aptera. His evenings would be idled away sitting on the inner harbour wall where *kareklás* weavers repaired brooms and wicker chairs with armloads of fresh-cut rushes. He'd watch the shoals of olive-scaled fish drift on the current, tracing a pattern across the sandy bottom, their lazy turns catching the setting sun and flashing here and there like the blade of a falling knife. After nightfall he'd walk home along pitch-black lanes, through pools of lamp light and cooking smells, the day's warm glow still radiating off south-facing walls.

Foreigners in Crete followed the advance of aeronautics through the rivalry of the occupying nations. First Captain Tupper might have passed around a dog-eared copy of the *Daily Mail* which reported on the Wright Flyer winning the Coupe Michelin at Auvours. Not to be outdone, French officers thrust upon visitors crisp copies of *L'Illustration* announcing Henri Farman's completion of the world's first cross-country flight. Farman had travelled twenty-five kilometres in a Voisin biplane from Bouy to Reims.

'*C'est le Triomphe de l'Ecole Française d'Aviation*', trumpeted *Le Matin* and no Frenchman would dispute the claim.

Next the Italian contingent boasted of the 'ascension' of the poet D'Annunzio at the Brescia air meeting, as announced by *Corriere della Sera*. It was appropriate, they pointed out, that while soldiers and politicians were carried aloft in northern Europe it was artists who took to the air in Italy. 'The new civilization, the new men, the new skies,' declared *Il Poeta*, likening the ecstasy of his first flight and return to earth to an orgasm. '*Una voluttà troncata.*'

Like most Europeans and Americans, the aviator would have been swept up in the euphoria. With time on his hands and a knack for carpentry he'd pore over the detailed plans of flying machines in *L'Aérophile* and *Scientific American*. He'd study the Wright Flyer, questioning the need for two propellers and the absence of longitudinal stability. He'd ponder the sturdy Antoinette's tendency towards 'trampling'. I pictured him making notes and drawings at a *kafeneion* table, his head buried in a report on the latest *Grande Semaine de l'Aviation*.

He appeared to me as a composite of authentic aviators: a quiet man of few words with the 'extraordinary keen, observant, hawk eyes' of Wilbur Wright; a determined glance and aquiline nose like Louis Blériot; a cheerful eagerness reminiscent of Eugène Lefebvre who was killed while making a low turn above the restaurant where he'd just lunched with friends. I created him tall and lithe, though not as slender as eight-stone Santos-Dumont, and able to run along a beach barely leaving marks on the sand. I dressed him in English tailored clothes like Hubert Latham. I gave him both an astonishing ability to remain calm, with no trace of emotion appearing on his phlegmatic face during moments of danger, and a moustache. I made him into a kind of personal god, at once veritable and wholly imagined, then suspended him between earth and sky, logic and inspiration, acting out a myth to help me to cope with life.

He was a man who did things rather than talked about them. Alone at the *kafeneion* table he dreamt of flying his own aeroplane, and – as he might have expressed it – 'gliding on the pinion of Hermes with Perseus and Icarus'. But the cost of a new machine would have been far beyond his means. According to the *Salon International de la Locomotion Aéro-*

nautique, a Blériot XI with a Gnôme engine would have set him back $5,000, twice as much as a luxury car.

I cannot tell for certain what effect Arthur Evans' discoveries would have had upon him, other than to bolster his determination. After all, the archaeologist had given substance to the labyrinth, so could an aviator not re-enact its architect's flight?

Then, one day on a tennis court or during a walk in the hills beyond Aptera, Captain Tupper might have given him a copy of *Popular Mechanics*. The magazine, though not often read in Europe, would have found its way aboard the *Venus* from a passing American cruiser. Perhaps it was the issue in which Santos-Dumont announced that the Demoiselle plans were now placed 'at the disposal of the world'. The working drawings of 'the smallest flyer ever built' could be bought for $5 (price postage paid), the very same set which was sent to me almost a century later.

'This machine is better than any other which has ever been built,' wrote the magazine's editor, 'for those who wish to reach results with the least possible expense and with a minimum of experimenting.'

All my aviator needed was time, lengths of bamboo, bicycle tubing and a place to build. Which is how I came to picture him walking up the hill from Vrysses, along the rim of the olive green basin to the village where Yióryio, Polystelios, mad Ulysses and I would one day live.

20. Something Old, Something New

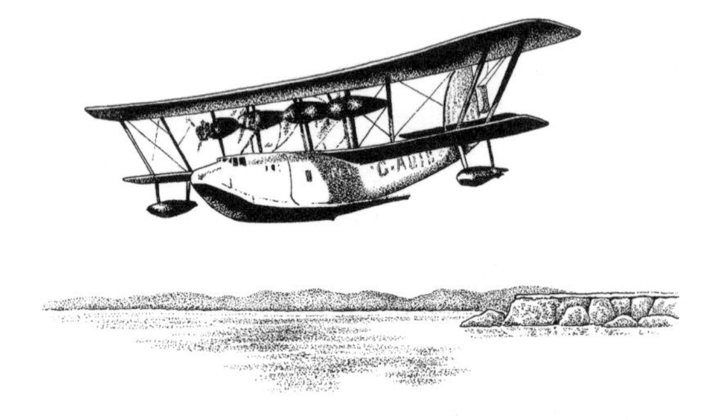

Ariadne gave me the number of her cousin. Her cousin's fiancé ran a moped rental company. I called him and by the end of our conversation about engines he had invited us to their wedding.

I had read about traditional Cretan wedding ceremonies with three thousand guests, volleys of rifle shots and a bridal bed made by virgins. I expected the manager of the Minotaur Moped Travel Agency to be a brawny groom. His bashful bride, Eleni, would be wrapped in yards of tulle and lace. The couple would be tearful as the *stephánia*, the white crowns linked by a ribbon, were placed on their heads. There would be lyre and lute players, *chaniótikos* dances with high leaps and thigh slapping. Money would be pinned to the newlyweds' clothes and every villager would have a seat at the table. Or so I imagined. We were unprepared for how customs had changed in less traditional parts of the island.

Ariadne drove us east to Ayios Nikolaos. Katrin wanted to

buy flowers for the bride and we stopped at the town's newest florist. Aristea, the effervescent proprietor, was a breathy, curvaceous acrobat wearing a clingy keyhole top, mauve boot-leg Lycra trousers and no knickers.

'Tell me all about the bride,' asked Aristea, leaping up from her game of solitaire. 'Is she young?'

'She is young,' confirmed Ariadne.

'I know who she is,' she guessed, flicking dyed blonde hair out of her hazel eyes. 'I know her. She is like me. I make the bouquet for me. You understand? With all my heart.'

Aristea skipped on her soles – not her toes – from counter to cooler, mixing pink roses with stalks of bamboo, deep purple leaves and gypsophila. Her scissors swept through tissue and coloured netting. She swathed the bouquet in giant folds of pink and purple rice paper, crimped and stapled tissue bows, then looped strips of colour between them. Her trousers were so tight that we could see her muscles moving down her buttocks and legs.

'These are butterflies. This is a rose. And this . . . I don't know what this is but it is fantastic.'

'That's enough green,' said Katrin.

'Not enough,' she said without hesitation, adding more eucalyptus.

In five minutes she created a plump, kaleidoscopic confection. 'Good better now best.' She waved away every suggestion and tentative look. 'I know what your friend prefers,' she assured us.

Ayios Nikolaos was a pastel-tinted, acacia-shaded harbour town and the capital of Lassíthi province. According to the local tourist office 'Ag Nik' had slumbered in obscurity until discovered by Melina Mercouri and Walt Disney. In the early 1960s the jet set made it a chic hideaway. One summer Karl Flick, the chairman of Mercedes-Benz, commandeered half a hotel for his entourage of bodyguards and fitness instructors. He only moved out when a Saudi oil minister sailed into the bay on his personal ship and rented the rest of the hotel. Package holiday-makers followed, spawning concrete-and-sunblock development, trampling underfoot the town's sleepy past.

The guidebooks maintained that tourism hadn't ruined Ayios Nikolaos, a claim which made Aristea look up from her cash till and laugh out loud. Yet beside her trendy florist's shop, fishermen bent their knees and tilted their heads, as they had done for generations, to sail under the low bridge into the inner harbour. Yachts bobbed alongside trawlers, with pots of basil in their wheelhouses and icons of St Nicholas, patron saint of mariners and the progeny of Poseidon. Oleanders still leant over the steep shore rocks. Old men played *tavli* in the *kafeneions*. Pigeons dozed on telephone lines.

We didn't stay long in town, or pause at the 'Twins . . . Taste That Wins' takeaway, but wound north over the arid hill towards Eleni's hotel.

'Crete will be the new Monaco,' announced Stelios, the moped king, to the early arrivals in the beachside piano bar. The wedding was due to take place the day after next by the tennis court. 'The Germans and Swiss are buying land. I am buying land. It is a golden opportunity.'

Stelios was not a tall man. He had neither bold eyes nor a chiselled chin. His teeth were small and imperfect. And he had no engine for me. But he blazed with such fiery optimism that sparks all but crackled off his new linen suit. So rather than stew through the weekend I let the Woodhopper slip from my mind.

'Hotels are theatres that take money from tired people.

Every fourteen days the actors change but the director,' he declaimed, striking a theatrical pose, 'stays the same.'

Eleni gazed at her fiancé through clear almond eyes, stroking his arm with slender fingers. She was the hotel's long-limbed, front-desk receptionist. Tiny beauty spots freckled her smooth, pale skin. She wore no lipstick and only a flick of lilac eyeshadow. She was proud, of course – I never met a Cretan who wasn't – and in her soft blue uniform she seemed to fill far more space than her petite figure occupied. She knew a dozen Cavafy poems by heart, had read Kazantzakis and spoke fluent German, English and Russian.

Stelios, on the other hand, had studied the writings of Conrad Hilton and Bill Gates. Instead of two-stroke oil he smelt of Givenchy.

'Every day when I see this place I am reborn,' he said, gesturing across Mirabello Bay towards the far mountains of Sitia. 'Can you say that? To be reborn?'

I assured him that it could be said.

'Are you a Cretan?' asked Katrin.

'I was born in Athens,' he replied. 'I came to Crete because I didn't want to waste my life. Athens is a beautiful city but no place to bring up children. Here I started my agency, met Eleni, now we marry and I will build my hotel. There!'

Stelios pointed over the water towards a last stretch of virgin coastline, its reflection shimmering in a silver sea.

'It is useful that Eleni's uncle is the town planner,' he added.

Fifty years ago there was no road to Elounda. The peninsula was so poor that it was said to produce only rocks. Visitors would come ashore and, after enjoying the local hospitality, see children fight for their discarded watermelon rinds. Now clusters of white concrete hotels clung to the shore like limpets and jet boats drove the fishermen off their fishing grounds.

'And you, Eleni?' I asked. 'Where are you from?'

'Kourounes,' she said, brushing a tendril of henna hair back from her high forehead. Her village's name meant the crown atop the hill. 'Just over that mountain.'

As dusk settled on Elounda we wandered down from the main hotel, past Venetian door frames beyond which lay private villas furnished with antiques, each with a walled garden, swimming pool and unobstructed view of the sea. The meandering lane descended through gardens of *almirikia* and carob, past three restaurants and a new Orthodox chapel, built at the same time as the resort and complete with Byzantine icons. Here and there paying guests lingered, listening to the soft whisper of the waves, anchoring themselves in deep armchairs, reluctant to leave the stage and break the holiday spell.

It was still Crete's low season and the hotel owners – as their gift to Eleni – were giving empty suites to her wedding guests. Katrin and I lucked out, finding ourselves in a small bungalow with steps sweeping down from the pool to a private rocky landing. As soon as the porter closed the door, we dropped our clothes and plunged into the glassy water. The cold was shocking and we swam hard to warm up, then floated on our backs suspended between the first evening stars and beds of clicking mussel-shells on the sea floor.

Beneath Elounda's sandbanks lay an ancient city dating back to Minoan times. Oloús had been built at the spot where Minos had chased the nymph Britomartis into the sea. Daedalus had carved a statue to the 'sweet virgin' which was reputed to be so lifelike that it could speak. Word of its marvel spread across the Classical world and the city flourished, growing to thirty thousand inhabitants after the Dorian invasions. Oloús traded with the eastern Mediterranean and the Ionian ports and, because of its sheltered, strategic location, thrived during the relative stability of the Roman and first Byzantine periods. But after Crete fell to the Andalusian Arabs in 824, Oloús' coastal location became a liability. The town was vulnerable to pirate attacks and its population retreated into the interior, risking a return to the plain only to plough, sow and harvest.

After the Fourth Crusade in 1204 the Venetians sailed their galleys into the bay, naming it Mirabello, or 'lovely view', and pilfering deserted Oloús' stone to erect two huge fortresses. They built salt pans in the shallows and, as the eastern end of the island gradually sank, the city's remaining foundations vanished beneath the waves. The land was so poor that the Turks took little interest in the area during their occupation. As a result Cretans wishing to escape Ottoman authority migrated to the peninsula. Eleni's forbears could have reached Kourounes in such a manner, seeking sanctuary after the disastrous 1770 uprising, when the rebel leader Dhaskaloyiánnis was skinned alive by his captors, or a century later following the destruction by the Turks of the Arkádhi monastery. Not that anyone in her family remembered the facts, and in the hotel's luxurious dining room no one seemed to care.

'It won't be a traditional wedding,' said Stelios, tucking into his starter of Parma ham with Cretan bananas. 'We're having a best woman rather than a best man.'

'And there will be no bride's bed,' Eleni explained.

The custom had been for the nuptial bed to be made with new linen, covered with flower petals and displaying the bride's trousseau – her life's work in needlepoint. Villagers passed through the room and threw gold coins onto the bed.

'Though I'd not object to the gold,' said Stelios.

'But the bed has to be made by virgins,' Eleni pointed out with a fleeting, sentimental smile.

'And there aren't any left on Crete,' said Ariadne.

'A few years ago my aunt asked me to help my younger cousins to make a bride's bed. I had to excuse myself and she was so shocked.'

'I read that village women used to lock themselves in the bride's house,' said Katrin, 'while the men stood outside singing, encouraging her to come out. The women inside sang back her refusal.'

'Nowadays people tend to meet on line,' teased Ariadne.

'Stelios and I met through a chat room,' laughed Eleni, wrapping an arm around Ariadne's shoulder, 'and he proposed in a cybercafé.'

'How about guns at the wedding?' I asked them. 'Will there be any shooting?'

At Anissari we had heard gunshots – and once a machine-gun volley – echo up the valley from Vrysses' churches every Saturday since Easter.

'Do we have guns?' Eleni said to Stelios, as if asking if he'd remembered the canapés. 'We must have at least one gun.'

'We know a man who makes Cruise missiles,' volunteered Ariadne. 'I could ask him to loan us one.'

We ate swordfish marinated in olive oil and lemon, washed down by a light Sitian white wine. Each year the hotel's manager sent two or three young chefs to apprentice at Le Manoir aux Quat' Saisons in Oxfordshire. I proposed a toast to the bride, adding as a pleasantry, 'I'm sure that you will be very happy.'

'I'm not so sure,' replied Eleni with a twinkle in her eye. 'So many marriages end in divorce these days. Let's just say I'm optimistic.'

'See,' Ariadne said to me, 'a real Cretan girl.'

*

The next morning Ariadne offered to help Eleni with final arrangements, including the arming of half a dozen members of the congregation, and in a gesture of solidarity they invited Katrin to join them. Stelios kept clear of the preparations and suggested instead driving me around the peninsula before meeting the women for a late lunch. We stopped first to look at his property, a mile beyond the sign 'Public Sandy Beach Sunken City'.

'The children's pool goes here,' he said as we picked our way along the rocky shoreline.

The morning was mirror bright, white sea reflecting white sky, and the mass of mountains invisible beyond the silver sheen of light. I'd left my dark glasses in the car.

'The conference hall will be built on this hill,' he continued, adding that corporate clients were essential to boost off-season trade. 'The top comes off the hill, of course.'

The beach rock ledges were razor-sharp and we needed to watch our footing. Rock pools along the shore teemed with crabs and bright pink algae. Stelios planned to truck in sand from the far Pyrgiótissa coast.

'Thirty-two apartments, twelve deluxe villas and a water sport centre,' he said in expansive mood. 'In season my target market is active twenty-five- to forty-five-year-olds. No one else in Elounda offers parasailing or banana boat rides.'

'Are those ruins there?' I asked. I'd spotted an edge of cut stone below the water's surface.

'Just an old building,' he said. 'It's way beyond the property line so the archaeological department isn't worried.'

Oloús.

I asked Stelios if Crete was being ruined by tourists. Or by Greeks.

'I feel pity for tourists,' he said, shaking his head. 'They live in countries with no sun and work like dogs to pay for two weeks' freedom.'

Before I could speak, he went on:

'The sunshine makes them randy, of course. Which was a problem at first. Greeks did not feel equal to tourists. How can you feel equal if you are a shepherd one day and a waiter the next? Swedish girls came here, made big cow eyes and we slept with them. We were the most macho guys in the world.

'Then AIDS arrived and – like always in Greece – we reacted in extremes. If we don't understand something we ban it. Overnight we had to *avoid* having sex with the tourists. Men – especially those working in hotels – said to each other, "What can I do? I can't screw her. She travels. She's probably slept with half of Africa. I have to think of my family's name. My wife. My kids." Then they took a breath and said, "But did you see her body? What breasts! What hands! . . ."'

We drove north from Elounda alongside the Gulf of Korfos, a natural harbour sheltered from the winds and known by locals as 'the lake'. At its head on a shelf of rock was Spinalonga. The treeless island had been fortified by the Venetians in 1579. Its monumental battlements protected the gulf and only fell to the Turks fifty years after the rest of Crete had surrendered.

At the start of the twentieth century the Christian Cretan government declared that the island would be a leper colony, in part to drive out the Muslim families who had by then inhabited it for two hundred years. The sick were shipped in from Heraklion, marched through the gate of sadness and left to fend for themselves on the narrow streets. As no new facilities were provided, the lepers took over the abandoned houses. Over the years a rough, despairing community took shape beneath the ramparts with a lane of small shops, a barber, cobbler and four *raki* bars. In the 1930s the sick went on strike to draw attention to the appalling conditions, banishing doctors, hanging great black flags from the turrets and ringing their chapel's bell day and night. Inmates often tried to escape from the island, especially if their condition condemned them to a life's sentence. One woman was caught when the crew of the boat noticed a burning smell. Her infected leg had been resting against the boat's exhaust, singeing her flesh, but as the leprosy had destroyed her nerve cells she'd felt nothing.

The colony was only closed in 1957, long after the discovery of drugs to control the disease. Twenty years later the tourist authority began to restore the fortress. Spring flowers now grew around the bone yard. The women who manned the entry booth continued to bring their drinking water from the mainland.

Beyond Spinalonga the road climbed above a wild coast, the hairpin bends skirting stone rivers of scree which washed down the cliffs. At a windy promontory we turned away from the sea into Eleni's battered land.

'The people here are mean,' said Stelios, 'because they never had nothing.'

The spartan and desolate peninsula had been moulded by the wind. There was no hint of working farms, tourism or human life. I felt as if we'd reached the edge of Europe.

'Many ships were wrecked on this coast, Pontius Pilate among them on his journey back from Jerusalem,' he went on. 'That's why Cretans call this place Aforismenos: cursed, windy. It is a bad word.'

It was also one of the driest places on Crete, so impoverished that its people had developed no tradition of hospitality.

Eleni had told Stelios that once there had been eighty-five villages in Aforismenos, none bigger than six or seven houses.

'In the whole place there is not one spring for fresh water. Rain falls on the rocks and runs straight into the sea.'

I saw the sharp line of a wall against the sky, the remains of a solitary croft. Around it prickly oaks and cypresses were scoured into wiry canopies. It was no wonder that the Turks had stayed away and fugitives found safety here.

'The tradition of this land is to protect the soil,' said Stelios.

The dry hillsides were covered with stubby walls, built low not to pen sheep but rather to prevent erosion. Stone paths wound across the landscape. Once almonds had been Aforismenos' cash crop. A kilo had been worth one and a half times the same weight of meat. Then, in 1937, the trees had been devastated by disease. Most of those that survived were chopped down for firewood in the war.

We drove through sparse hamlets called Agia Sofia and Dilakos. Makrigennisa was a collection of tumbledown buildings hidden at the bottom of a steep valley. Here there were sheep, huddled out of the wind in a hollow. Eleni had told Stelios that fifty people had lived in Makrigennisa a century ago. Today only an eighty-year-old woman remained.

'Eleni brought me here to meet her. I asked her, "Don't you feel lonely?"

' "I had my husband with me for forty years," she replied. In a few more she too will be dead.'

And another village will vanish.

In the lee of the wind grew wild asparagus, beloved by goats, and poisonous purple *dragondia*, the 'female dragon'. Here too thrived rabbits and birds, so isolated in this place that they went unhunted.

'You know here is small,' said Stelios as we drove further inland. 'Small minds. No dreams. Even your neighbour, if you smile at him, he ignores you. You think, "Is it me? Do I smell?" The people speak only in Cretan language. And the crazy thing is Eleni would like to live in this desert. Not in Elounda where she can speak English, French, German . . .'

'. . . and Greek?'

'Even a little Greek,' he admitted with his infectious smile.

We passed through an empty village called Nofalioas. Its name derived from *omfalós*, the word for umbilical cord, the centre of the body. Stelios turned up a single-track lane into tiny Kourounes. Eleni was born and grew up in this hamlet, the crown atop the hill at the centre of Aforismenos. No one was out on its single street and Stelios didn't want to stop.

'Do her parents still live here?' I asked.

'They moved to Neapoli,' he said. New Town. 'Nobody lives here any more. This is a place of the dead.'

Aforismenos was an open museum to a hard, lost life, beautiful in its arid austerity, though not for anyone looking for a sandy cove to pitch his beach umbrella. Ten miles away holiday-makers rode hot rings and drank piña coladas at the beach bar. We returned to that world almost as suddenly as we had left it, dipping back beneath the lip of the hill, out of the wind and down to our rendezvous at Elounda.

'How was the drive?' asked Eleni.

We told her our reactions to Aforismenos. If she really did want to stay in her village, tending the family's last almond trees, she didn't mention it. At the hotel she earned three times the salary and, in any case, in Greece it was the husband who chose where to live. I added that I was interested to see the submerged ruins.

'That's Oloús,' she replied. 'Isn't it the perfect place for the hotel?'

We sat in the sun outside Evangelia's Snack Bar looking across the water towards Spinalonga. Evangelia, a wafer-thin woman in leather trousers, was repainting a florid pizza on her summer menu board. She put down her brush, took our order then sauntered off to buy bread and fish. The village postmaster joined her at the mini-market, leaving stamps and money on the counter for customers to serve themselves. Ranks of rental cars – most with hubcaps missing – lined the main square, washed and ready for the season. An old Mercedes taxi drifted past them looking for conversation or a fare.

We ate grilled grouper and sea bass, caught an hour before by a fisherman filling out his pools' form at the next table. Eleni picked at the fishes' heads rather than their bodies.

'It's my favourite part,' she said.

'In Greek we say "The fish smells from the head",' said Stelios as he snatched the bill from my hand. 'Meaning that the head is the first part of a fish to rot. For example, if a hotel's management is no good, the staff will be bad.'

'Money and fish, you have to eat them both fresh.'

After the meal Eleni convinced Stelios that his help was needed at the hotel finalizing the seating plan. Ariadne led Katrin and me beyond the Kalapso Taverna to the church of Agia Triada. A dusty motorcyclist glided by exhaling cigarette smoke and looking like a steam-powered Minotaur. In the graveyard white plastic roses decorated marble stones. At the head of each tomb sat glass boxes like bathroom cabinets, containing a photograph of the deceased, an oil lamp and – more often than not – a swatch of fishing net.

An eight-foot-square concrete grave, by far the largest in the cemetery, stood beneath a feathery maritime pine. It had neither glass box nor flowers. Its simple marble plaque read, 'In loving memory of Lieut. Richard Glen Wilson Dickson Fifth Battalion Royal Tank Corps and Thomas Alexander Cecil (Whimmie) Forbes late 16th Lancers and Indian Police who lost their lives as a result of a flying accident in Mirabello Harbour 22nd August 1936.'

Through the 1930s Imperial Airways flying boats operated across the Mediterranean, setting down in the Gulf of Korfus to refuel. Outbound from Europe to Africa the four-engine

Short S17 biplanes – carrying sixteen passengers at 105 mph – circled above cursed Aforismenos to line up for a final approach over Spinalonga. A sack of table-tennis balls was emptied above the bay to enable the pilot to see the surface of the crystal-clear water. Then the flying boat would touch down and on one engine tack towards Elounda Harbour. At the forward hatch the co-pilot threw the bow rope to the MV *Imperia*'s six-man crew – swarthy locals in knitted acorn hats and company jerseys – waiting at the refuelling pier. Passengers stepped ashore to stretch their legs, walking as far as the carob processing factory, bending down to pet the *Imperia*'s mascot terrier. Once or twice they posed for photographs with the station's steward, tall in his crisp white uniform and air of importance, before reboarding their flight to Alexandria.

The lepers would have watched the silver bird too. From their island prison they saw it lift off from the tranquil waters that lay between them and the outside world. They could see it as it banked around the blue Sitia mountains and flew out of sight. They also witnessed their first aeroplane crash. In August 1936 on its final approach, 'Scipio' G-ABFA nose-dived into the water. In error the wireless operator had moved the tailplane trim to full nose down position. The flying boat's hull was torn open on impact, the wings buckled and both Richard Dickson and 'Whimmie' Forbes were killed. That night the lepers of Spinalonga hung a black banner from the Venetian turret and for months the wreck lay on the sandy bottom. Subsequent Imperial passengers had the unnerving experience of landing over the sunken plane.

'The water is so clear,' said Ariadne, when we paused at the Nautilus Bar beside the old refuelling pier. 'It's a beautiful place to fly.'

'I don't think so,' I told her.

★

On a wedding day – as on the first day of the year – a pomegranate smashed against the threshold brings the promise of prosperity, anticipating the return of Persephone, goddess of death and the living earth. But no one wanted to make extra work for Housekeeping.

Almost every Greek couple gets married in church, not because civil weddings are discouraged, but because a man or woman married outside the church might later be prevented by the clergy from being buried in consecrated ground. So I hoped that the ceremony would be held at nearby Panayía Kirá, the Byzantine church with some of the finest frescos in Crete. On its medieval walls holy-minded saints clutched crosses and unrolled psalms with long, slender fingers like Eleni's. Gregory the Theologist wore a catfish beard not unlike the Winged Priest's. But Eleni's boss had offered her the elegant chapel above the Yacht Club restaurant and Stelios had insisted that she accept.

'They are influential people,' he whispered to me as their guests – brothers, aunts, sous-chefs and the manager of the local Hertz franchise – idled into the concrete nave, a space bare of decoration but for an altar of antique icons imported from the mainland.

'I hate cliché,' said Ariadne, introducing us to another of her relations, 'but I'm glad the bride is late.'

'The woman's last attempt to dominate the male,' shrugged Stelios, waiting at the chapel door holding a bouquet. 'Excuse me but I'm a chauvinist.'

The *koumbára*, or best woman, was the hotel's food and beverage manager. 'The secret to a successful marriage,' she advised Stelios, 'is to want what you have.'

'I don't have forever,' he replied and looked at his watch.

Four severe priests – among them Gregory the Theologist – glided through the throng, unsnapping their briefcases behind the altar to extract white cassocks and breath mints. Beside them Eleni's boss lit the candles, great white phalluses swathed in fabric, and the staff passed around glasses of *tsikoudiá* to lubricate conversation.

'*Ela!* Here she comes.'

Eleni's arrival was announced by warm applause. She stepped out of the car wearing a low-cut, home-made dress, and stood on tiptoes to apologize across the tops of the heads to Stelios. He rolled his eyes. They met at the door, kissed and entered the church.

All the congregation squeezed in behind them, or hurried around to the side door, encircling the couple, not turning their backs on the icons. Cameras flashed, babies cried and the bridesmaid's straps fell off her shoulders. Eleni's grandmother used her walking stick to prise her way to the front of the crowd. No one stopped talking. I could barely hear the chanted service above the chat and shuffle of the congregation.

At first the bride appeared self-conscious, even bored. Then, as the sauntering pageant became more physical, she was in turn amused, moved and entranced. The priests jockeyed for position, the Bible was kissed, wine drunk, rings and the simple flower *stefania* – garlands – touched to head and heart of man and woman. At the back of the chapel Stelios' business partner – a moped importer – answered his mobile phone.

'*Ela, Dimitri! Pou eisai, maláca?*' Where are you, you wanker?

The air grew heavy with incense and wax. My back warmed against a well of candles. No one cried or laughed and the unmystical service left me unmoved, which surprised me, though not the Hertz agent who popped away for an hour in order to deliver a car.

The priests told Eleni to fear her husband. In a customary act of defiance she tried to step on his foot. He stamped on her toes. Then they were married. Gregory the Theologist smiled and said, 'Dance, Isaiah, dance', leading husband and wife, joined by God and the ribbon between their crowns, around the altar three times. The guests filtered in from the cool courtyard, throwing rice and flower petals, pushing forward to kiss bride and groom. Eleni kissed the hands of the priests.

At the door we were given sugared almonds.

'Women are meant to put these under their pillow so they'll dream of their true love,' Ariadne told us. 'All I ever see is Apostoli's father.'

In good society brides no longer pin money to their bodices; rather, guests make a show of leaving plump white envelopes on silver salvers.

'Honey means sweet life,' explained the best woman, offering us honey and walnuts, the Minoan symbols of fertility, on a disposable plastic spoon. 'It also preserves the sweetness of the marriage.'

Then she blew out the phallic candles, snipped off the blackened wicks and gave them to the bride.

'So no one can take them and make sorcery.'

The staff had arranged long tables on the tennis court and guests took their seats to eat Eleni's mother's *paidhákia* – lamb chops cut from the ribs – and to drink her father's wine.

'What, no cake?' complained a disgruntled aunt from Kalamata. 'You are breaking too many traditions today.' In fact it was one tradition which Eleni and Stelios weren't breaking, wedding cakes being a recent innovation from northern Europe.

We sat with the aunt, the moped importer and the hotel's Austrian watersport coach.

'Have you tried accuweather.com?' he asked me. 'It's the best forecaster on the web.'

On a makeshift stage the musicians smoked and tuned their lyras while waiters served one hundred in a sitting, twisting between the tables with enormous platters of macaroni, *spanakópita* and *mizithrópites* pies.

The Austrian declined an offer of wine. He preferred scotch.

'Cretans think it insulting to refuse hospitality,' I reminded him.

'If their friendship is based on my accepting their drink then I should find other friends,' he said, dismissing the custom out of hand.

He explained that he taught skiing in the Alps in the winter and came to Crete for the summer, bringing with him his German-language video library.

'This year I've bought a satellite dish so I can watch real television.' He added, 'I miss civilization when I'm here.'

I suggested that his stay would be more rewarding if he respected local traditions.

'All summer the Greeks subjugate their culture for the tourists. They throw away Coke cans and throw up concrete and make everything so ugly. They answer the same questions again and again: "Yes, it is hot today – Yes, you can swim topless here." Why should I respect such banality?'

'I want to talk to you about this Meggitt engine,' said Ariadne, leaving her cousins and taking the chair beside Katrin.

I turned away from the Austrian.

'We have to think how best to get it here.'

'DHL?' I suggested.

She laughed at my naiveté. 'It is a military engine,' she said.

'So we can't get it through Greek Customs?'

'Not Customs,' she corrected. 'There are no Customs between European countries. Let's call it Inspection.'

'Then how do we get it through Inspection?'

'Ship it direct to me,' said Ariadne. 'In my name. With my address. I will arrange everything.'

I pointed out that the package would be much too large to slip unnoticed into the country.

'Last month I ordered tank tracks,' she said with satisfaction.

'Tank tracks?'

'For Socrates' digging machine. The shipper faxed me the waybill and flight number. I went to the airport to collect them. I saw the aircraft land, then take off again. No tracks were unloaded. The airline couldn't find them. I spoke to the manager who said he had no record of them.'

'"Here's the waybill number," I told him. "That was the aircraft. Look on your computer and find out where they are."

'So he called Heraklion and Athens but they had disappeared.

'"What are you?" I shouted at the man. "You are the General Manager of Cargo and you can't find my stuff? Why don't you quit your job and give someone else a chance?"'

All of our table was now listening, relishing Ariadne's tale.

'There was another woman there too, a tiny woman who had flown from Athens with her dog in a cage in the hold. And they had lost the dog. She was screaming at the manager, "You have lost my dog. Have you no heart? She is a life. She has a soul. She cost me money. And you have lost her." The woman was hysterical. "I'll kill myself if you don't find my dog. I'll kill myself and come back as a ghost and kill you."'

'Then Athens rang back and told the manager that the tank tracks and the dog had been left on the aircraft and were now back in Athens.'

'"Why didn't your men unload the tank tracks?" I asked the manager.

'"Maybe they didn't see them," he answered.

'"They didn't see two tanks tracks – weighing two hundred kilos – and a dog barking in its cage? Are you crazy?"

'So we waited for the ten o'clock flight,' Ariadne went on, beaming, 'the last flight of the day. It came, and the tank tracks were on it as well as the dog. "Piss everywhere," the woman told her dog. "Piss on the marble floor and everyone here."'

'Are you suggesting that this is a safe way for me to import the engine?' I asked Ariadne.

'Of course. If it is addressed to me I will get it for you. The Inspector will see it . . .'

'. . . in its for-military-use-only packaging.'

'Yes. The Inspector will see it and he will ask me, "This looks valuable. What's it for?"

'"My cousin's rotavator," I will say.

'"How much is it worth?"

'"This piece of junk? Nothing," I'll say.

'"One hundred thousand drachmas?" he'll say.

'"Are you kidding? It's worthless I tell you. My brother will only use it in the garden."

'"Fifty thousand drachmas then."

'"I wouldn't give you one drachma for this old garbage," I'll tell him. "Keep this shit. You can throw it away."

'And this is where I'll have him,' explained Ariadne, basking in our attention. 'Because if I give it back to him he has a problem. He has to return it to England at his expense. So I'll say, "I'll give you something for coffee but I'll pay nothing for the machine." And then you'll have it.'

'Thank you,' I said. 'I think.'

'I don't do this for everyone, you know,' said Ariadne.

As we laughed the musicians began to play, the tune lifting rather than subduing the volume of conversation around us. No one was willing to give ground to the lute and lyras. The moped importer, who was in his cups, shouted above us all for the dancing to begin.

Hand in hand the newlyweds stepped onto the wooden dance floor.

'She moves like a Nereid,' cooed the importer, watching Eleni shake grains of rice out of her hair and cleavage. 'Like a *Playboy* bunny.'

Stelios took off his jacket and spun around, his arms spread out in an arc before him, defiant and proud. He laid his hand on Eleni's shoulder, she wrapped her arm around his and, watching each other's feet, they tapped the ground then kicked back. They dipped and rose, as if to overcome their bodies' weight. They stepped sideways, wrists twirling, eyes radiant. Their dance grew in circles and spirals as husband and wife warmed to the plaintive melody, finding their confidence and feet.

'*Opa opa*,' called out the importer, clapping his hands with the rhythm.

Eleni invited her father and new father-in-law into the dance. One by one all the male relatives joined them, linking together in the circle with outstretched arms. Table by table, family by family, the other guests followed, clasping hands with the dancers, wheeling, whirling, embracing expressions of community and continuity in the cyclic pattern.

'*Ela soú léo*,' cried Ariadne as the dance gathered momentum. Come on!

The instruments began to wail, in solo and in harmony, accelerating and played with an energy as concentrated as the smell of oregano. Their sound was insistent, athletic and magnetic. A waiter put down his tray to join the dance too, extending the spiral and metaphor, while fellow staff members began to stamp the earth until I thought the olives would fall from the trees.

The lead musician began to sing a rough wedding *rizitika tragoúdia*, or 'ballad from the roots', its sliding strings and bending notes echoing the East, especially when he ended it with the refrain, '*Aman! Aman!*' 'Mercy! Mercy!' in Turkish.

A second singer followed with a traditional *mandinada.*

'I've told you once, I've told you twice, I've told you three times, haven't I? . . .'

At the next table the men picked up the words.

'. . . You must no longer wiggle your rump or shake your breasts . . .'

Our table replied in a horrible screech, each man – and Ariadne – singing in their own pitch.

'. . . Driving mad the young men and the heroes . . . and the priest!'

Even Gregory the Theologist laughed when Stelios slapped his wife's bottom, though he stopped banging the table when she in turn patted her husband's backside.

'That girl will have to fight,' Ariadne yelled to me.

The *mandinades* were bawdy love songs, embracing birth, baptism, marriage and death. The couplets told stories of dark-eyed girls and heroes as tall as cypress trees filled with Homeric courage. Their world was a deceitful place where Charos the reaper stood ready to scythe away their beloved.

'Wake up, my love, the day is already dawning; give your fine body to me and the clean air,' crooned the moped importer with tears in his eyes, until drowned out by others singing their own repertoire.

'His mother will die who sings to the end,' the Kalamata aunt warned him.

'When I die bury me with my mobile,' warbled the lyra player, his *askités* inspired with a modern twist, 'but not too deep or it will receive no signal.'

Between the *mandinades* the music grew wilder, accelerating to a frenzied pace. Eleni stepped round and round in her voluminous skirt then broke away from the ring, leaving Stelios to dance with her younger sister. She didn't sit down but moved from table to table talking to her guests.

'The family will expect a baby soon,' the aunt informed her when she reached our table.

'In theory they will have one,' said Eleni.

'Ariadne told me that you found some guns,' I said to her.

'When the men start shooting,' she advised us, 'don't be scared. They usually know what they are doing.'

'Usually?' I asked

'Usually.'

Ariadne, who had wandered off in search of ice cream, returned to the table and the subject of the Woodhopper. I'd started to doubt whether it was a good idea to ask Westlake-Toms to ship the engine to someone other than myself. It seemed unappreciative, even rude.

'I think I should collect the engine myself,' I said, planning to bring it back as checked luggage.

'In that case pack it in a tatty suitcase or it will be stolen,' said Ariadne. 'If the baggage handlers in Athens see Samsonite they will have a look. If they see only a sack they'll say, "Leave the poor bugger alone."'

'If you bring it yourself watch out for Hercules and Odysseus,' Stelios said, joining us during a break in the dancing.

'The old heroes?'

'Our new robots,' he replied.

Earlier that month two 'sniffer' robots had started work at Athens airport where I would have to change planes. The Odysseus robot identified suspect packages and handed them to Hercules who placed them on a special platform and destroyed them.

'I think Odysseus will smell your petrol engine.'

On the soles of Eleni's shoes were written the names of all her single girlfriends. The name which hadn't been rubbed away by the end of evening would be the next woman to be married. But when Eleni and her aunt looked all the names had been erased.

'This is a good time for us,' said Stelios with loud optimism, joining us to wrap his arm around the moped importer. 'More tourists than ever came last summer and my franchise is making a killing. We are no more an island of farmers.'

I thought of the shepherds of Anissari. Sheep had been part of the landscape for thousands of years but in the last decade had become almost worthless. Their wool was burnt and their meat fell short of European Union export standards.

'So we'll be all right,' said Eleni, gazing at her husband, in love, confident, but with an unexpected edge of hesitancy in her voice.

'Old Crete is ruins,' said Stelios. 'Now come,' he said reaching for his bride's hand, 'let's dance.'

Under the stars in the lee of the luxury hotel the wedding celebrations went on all night. Rounds of family *kefalotyri* were served alongside platters of brie, imported from Slovakia. We drank more wine and Gordial Dry Gin, 'produced under licence by the General Chemical Laboratory, Athens'. Tree frogs croaked in the breaks between songs. Children fell asleep under the tables. The dance floor collapsed. The moped importer groped the bridesmaid in the chapel. The Austrian went to his room to watch a video.

Sometime after midnight the guns started firing, three

shotguns and an Ottoman flintlock which had belonged to Eleni's grandfather. Volley after volley resounded between the concrete walls and the far hills.

The next morning Katrin and I borrowed Ariadne's car to drive alone around the gulf. I wanted to climb into the hills to get a sense of the shape of the land.

We followed the curve of the bay eastwards, squeezed between a line of pylons and the sea. Golden marigold decked the fields and Judas trees splashed crimson over the hillsides. At Kavoussi where tour buses and Texaco tankers ground up the incline, we parked the car and picked our way along a broad, dry riverbed. An ancient, stone path struck uphill, weaving through the shade of olives and overhanging terraces. We followed it, sidestepping serpentine irrigation pipes, climbing back in time.

The path narrowed beyond the reach of village voices. The amplified patter of travelling salesmen fell behind us. A white chapel above us vanished behind a cliff. Dry leaves crackled like cornflakes under our feet. The undergrowth tore at our ankles then our knees. A single cypress marked a ruined farmhouse, its earth roof collapsed with its vines, its terraces abandoned to mountain sage. At the foot of a crumpled wall lay the hair and bones of a dead goat.

The chapel, once above us, reappeared from around a headland far below. Mirabello's bay opened up to the north, a turquoise crescent in the afternoon sun. To the east the bare, massed mountains huddled together like giants bent in conversation. Mount Kapsas tilted into the sea, a conical dune of scree. Above its shoulders circled eagles, their shadows falling across the path.

A jumble of stones marked Gourniá, the most complete of the Minoan towns. Its once pivotal position, controlling a narrow isthmus between the Cretan and Libyan seas, had made it an important trading centre. Alabaster from Egypt, emery from Naxos and lapis lazuli from Afghanistan had passed through its cramped lanes, en route between Asia, Africa and Europe. Some of the material had been worked into armlets and seal stones in the now-excavated workshops.

We walked over the cobbles, trying to sense a pattern in the ruins, working our imaginations until the scattered rocks found order and we discerned village streets and foundations. A flight of stone steps led up to the vanished homes of mud brick and plaster-daubed reed roofs. Grindstones lay where last used. We stepped up onto the main square, into the governor's quarter, over thresholds worn by Minoan feet. For all its bareness Gourniá felt familiar, even intimate, and not unlike Greek villages of fifty years ago. We pictured donkeys tramping up the alleys laden with timber for carpenters and copper for the smiths. I even fancied that I could hear laughter, or maybe the sound of crying, from the marriages once performed and deaths endured in the compact, industrious town.

The view hadn't much changed in the 3,500 years since Gourniá had died. The hill still commanded a majestic panorama. From its vantage point the Minoans, or a passing god, would have watched the Santorini tidal wave swell over the horizon to drown their civilization. Athenian triremes would have followed it into the bay, as well as Roman and Venetian galleys, corsair frigates and Turkish men-o'-war. A patient observer would have seen ancient Oloús sacked, sunk and dismantled to build the Spinalonga fortress. He or she might even have caught sight of Eleni's forefathers, planting their almond trees over the lip of poor Aforismenos, and the lepers being ferried to their island prison. The silver Imperial flying boats would have risen up into view too, flashing in the sun, banking over the ruins of the town to cross the island as the ancients had done, flying on to another world.

Beneath the carobs a hummingbird hawk moth fed on a spill of purple flowers. Above us a red-backed kite climbed on the thermals with no more than a lift and twist of its black wing tips.

21. Down to Earth

On May Day the air was so clear that I could almost touch the tips of the White Mountains. It was the day that Persephone emerged from the Underworld. Her spring flowers appeared on front doors. Drivers gathered asphodels and orchids to tie to windshields and bonnets in her honour. The seasons changed with a twist of the planet, stirring the villagers, sending young men to girlfriends' houses, revealing for a moment the invisible net which enfolded heaven and earth.

'What next?' asked Apostoli.

He and Ariadne had arrived late that morning, her first visit to the garage in three weeks. In spite of her professionalism, and a morning spent servicing an F-16 which had crashed into a tree after its pilot lit his afterburner while taxiing, she couldn't contain her excitement.

'I can't believe it,' she repeated over and over. 'I can't believe that you have done all this.'

She walked from wing tip to wing tip, running her fingers along the leading edge, drumming the fabric.

'I'm astonished by so many wires.'

'Too many wires,' grumbled Apostoli.

'Will you paint it?'

'Only the essentials,' I said.

Nothing that was not needed.

'We must have a big feast and baptize it,' said Ariadne. 'Woodhopper is not a beautiful name. It needs a human name that will inspire others to come close to it. You understand, your aircraft has a soul. I want to be its relative.'

I remembered finding an article written in 1908 about the Wright Flyer. 'It looked like the work of an amateur,' a reporter wrote in the *Daily Mail*. 'It was rough, almost make-shift. The two tiny seats on the lower plane seemed to be made out of biscuit boxes.'

'In my mind sticks the name of the wind: *ánemos*,' said Ariadne, cupping my elbow in her palm. 'We can call it *petás san ánemos*. Flying like the wind.'

I thought that she was about to recite a poem. Never had I known a machine to evoke such emotion in an engineer.

'I'll stick with Woodhopper,' I said. 'It's down to earth.'

'You know, you don't need an engine or airfield at all,' she told me. 'You just push a little and it will fly on its own. *Bah!* It lives. I say again, it lives.'

'I need the engine, Ariadne,' I assured her. 'I *have* to fly.'

She ran her hand down the boom, checking the tension on the wires. 'At the beginning I thought you were stupid; a writer wanting to be a craftsman. Now after what I see today everything seems possible. For sure you are crazy but without this sort of madness we cannot survive. Come, I buy you a drink,' she proposed, 'and we discuss one or two small modifications.'

We felt exhilarated walking to the *kafeneion* discussing the date of the maiden flight and the numbers to be invited to the airfield, wherever it might be. Ariadne promised to make a square steel sleeve to slide over the boom's nose and mount the Meggitt engine. It would be ready by Tuesday, the day I was booked to travel to England. On Wednesday she was scheduled to service a Hercules. On Thursday she would collect me at the airport, negotiate with 'Inspection' and bring me and the engine back to Anissari. I would fly on Saturday, all being well.

'The weekend is best because no one is looking at the radar,' she assured me. 'During the week the 340th Squadron flies sorties and they'll think you are the new Turkish invasion.'

'The Greek Air Force doesn't fly on weekends?'

'Come oooon . . .' she said. Meaning of course not. 'Now pay attention *pilóte*, you have to consider where you will fly.'

I understood her. It was her way – the Cretan way – of telling me that finding the airfield was my responsibility.

'We'll look tomorrow and on Monday,' I said.

'Ask around the village,' she suggested.

'Now is the time,' announced Yióryio as we entered the bar. Word – or at least our wave of elation – had preceded us. 'Today is day,' he said, pouring out six glasses of *tsikoudiá*.

We took a table in the sun as I whispered to Apostoli, 'What will you do?' He had told us that he never drank alcohol.

'I don't drink,' he confirmed, then added with assured logic, 'but I never say no to *tsikoudiá*.'

'This is special for liver,' declared Yióryio, toasting us and the aeroplane then emptying his glass.

As did Apostoli with the words, 'Good night.'

'You can never imagine anything better to do in Anissari than build an aircraft,' said Ariadne, still flushed with excitement. 'I am sure everyone on the island will create stories about you, jealous stories.'

The mention of stories caught Yióryio's attention.

'The mayor wants *panegyri*,' he said. A festival. 'To celebrate aeroplane.'

'A baptism,' said Ariadne.

'Can it wait until after the flight?' I asked.

'And if you not come back?' asked Yióryio.

Was he saying my death would deprive them of a party? I felt my temper rise. I was so close to the flight, so close to moving on with my life. I could hardly contain my anger.

'He will come back for sure,' said Ariadne, taking my hand to calm me. 'So the *panegyri* can wait.'

'But news cannot,' Yióryio replied, at odds with her. He was angling again to invite Cretan Television to the village. I didn't want the spread of news to compromise the flight by broadcasting my intention to break the law.

'I'll agree only to filming on the last day,' I said, knowing that any broadcast would be after the flight when the aircraft had been dismantled.

'*En táxi*,' Yióryio said, offering me his hand. Agreed. He seemed satisfied, refilling the glasses with his own brand of domestic aviation fuel. Until he added, 'But maybe they come tomorrow anyway?'

I didn't want to upset the villagers. I needed them, and not just for wine and eggs. They had helped me – or at least aspired to help me – find the garage and engine. And more. Their displeasure could ground me in an afternoon. But they were Greeks and, as I had come to learn, expected a payback. They wanted publicity for the village.

I had three days before flying to London and there was still work to do. Back in the garage Ariadne now saw the Woodhopper not as a thing of beauty – aesthetic and inspirational – but as a machine to be trimmed and balanced. She looked at its line and the relation of angles. She read the twist of the tail, the tension of wires, the lack of a smooth leading edge on the elevator. There had to be a four-degree positive angle of attack on the main wing. The wires, which Apostoli and I had set by eye, needed to be tautened. I saw the technicalities as a kind of metaphor for something that appeared to me to be finished, which, seen with different eyes, needed attention to detail.

'How will you start the Meggitt engine?' Ariadne asked.

'With a cord?' I guessed.

I had imagined that it would have an on–off switch.

'It's not a chain saw,' said Apostoli. 'You need an electric starter and I have one.'

'You do?' Ariadne asked her son.

'Of course,' he said. 'Although it would be better if Meggitt loaned you one of their own. Mine may not work.'

At first the alcohol fired us through the afternoon but its effect soon waned, taking our energy with it. Apostoli slumped on the workbench, stunned by the drink. He had sprained his arm in the gym and now hadn't the strength to tighten the Nico sleeves.

'My coach tells me that I have too much muscle for my body size,' he said. 'I can work no more today.'

Ariadne brought him a Coke but its caffeine failed to lift his spirits.

'I think I lie down now on the good earth and die,' he announced. 'Here on May Day I will say goodbye to you, friends and mother. Goodbye and good night.' He lowered his dark glasses over his eyes and promptly fell asleep.

On Saturday Apostoli did not come to work. The Swede had returned to Stockholm and his girlfriend felt neglected. She claimed that he never sent her text messages from the garage, which was true as he was forever on the phone to a third woman. She worked at the airport control tower and rang him every time an interesting aircraft came in to land. At least six times a day he'd cry 'Phantom F-4!' or 'E-2C Hawkeye!' from beneath the Woodhopper's wing.

Katrin too had been working less often with me. She did come when I needed her, which was most days, but sometimes our tension erupted into argument. Yet throughout she held her fears in check and blinked away a truth about which she could not allow herself to think.

She helped me to fasten the last wing wires, making openings in the fabric with the branding iron and threading the cable into the anchors. I lay on the floor pulling, swearing and developing calluses. Each new wire and adjustment affected the tension of its neighbours. Apostoli had wanted to attach turn-buckles to the looser wires. I worried that they would weaken the overall structure, especially as the only turn-buckles available in Vrysses were for vine cultivation.

Ulysses and Leftéri spent the morning with us, as usual, the madman sitting on his stool, the child in the pilot's seat. Together they soared off to imaginary destinations, Ulysses carrying the silent radio and Leftéri making his singing bird calls.

'Where do you fly to?' Katrin asked him.

'*Stón ouranó*,' he replied. Heaven.

'Do you know the story of Icarus?'

'I don't know it.'

'It happened a long time ago.'

'When did you see him?' he asked.

After the wires we worked on the tail, balancing half a dozen books on the elevator to correct its warp. Now only the control system, seat belt and moped wheels remained to be attached.

One of the tyres had a puncture and I proposed bringing

my bicycle repair kit back from England. Katrin looked at me with disapproval.

'You want me to buy a new inner tube?' I intuited, ever the perceptive husband.

'They're not so expensive.'

I needed to watch my budget. The parts for the aircraft had cost me almost £600 and now I had the unexpected expense of the London flight. But I figured I could stretch to a new inner tube.

'And you'll need to wear something thicker than those ancient Gap trousers.'

Across the square Yióryio let out his guinea fowl and leant in the doorway of the *kafeneion*. Farmers stopped in for *éna métrio*, their children came home on the school bus, their wives stooped to pick *hórta*, as they did every day. Sophia hung out her third load of washing then ran her daughters into town for an English lesson. Socrates the shepherd drove two dozen sheep between his fields, their bells jangling. His sulky teenage son flushed bleating stragglers from the verges and didn't return my wave. The church clock struck six, three minutes late, and we closed the garage door and set off to find an airfield.

22. Some Hope

Katrin and I planned to circumnavigate the island as far as roads allowed. At Skaleta on the northern coast the highway climbed onto a barren plateau overlooking the sea. Socrates had told us to look for a landing strip beyond the Cretan Farms' abattoir. We did and there wasn't one. We drove back and forth along the road, then asked at a Volkswagen showroom standing alone on a promontory high above the sea. No one knew of an airfield.

One suggestion down, five to go.

Next we backtracked to Gerani where, as advised by Kóstas, the New National Road petered out at a fluorescent Road Closed sign. We skirted it and rattled down a rough earth ramp onto a ribbon of new, black asphalt which sliced through the olives towards Kolimbari. Along its five-kilometre length there wasn't a single construction worker or crash barrier. Unfortunately there were deep hillside cuttings of raw stone into which I could slam the Woodhopper, after being electrocuted by any one of the dozens of overhead power lines. I might also be run down by the constant stream of traffic speeding along the unopened, unpoliced road. Kolimbari too was less than ideal.

Two down.

We turned south towards the old Luftwaffe runway at Timbaki, Crete's drabbest town. Beyond a dusty concrete sprawl and acres of polythene greenhouses was an airstrip, as Little Iánnis had promised. The location was ideal: at the edge of the Messari Plain, facing the sea, near to a hospital. But Iánnis hadn't mentioned the high security fence and armed military police at the gate. They didn't offer us a tour of the base.

The westbound road fell out of the mountains and descended so close to the shore that I lost my sense of proportion and distance. Then the tarmac swung back on itself and climbed again into the clouds. I was reminded of the Greek sailor of urban myth who asked, 'Are we sailing straight or is the coast crooked?' Sheer limestone cliffs rose a thousand feet above the Libyan Sea, pushing us inland into deserted, desolate valleys. Nothing seemed to live along this part of the south coast apart from mountain goats whose forms were silhouetted against the sky; bony, black and all but indistinguishable from the rocky ridge.

We tipped into Sfakia, the wild province which Romans, Venetians and Turks had all failed to subjugate. Fifteen parallel gorges ran into the sea within a distance of twenty miles, isolating villages from each other and the region from the world. No road reached Sfakia until the 1950s. The isolation bred a people of fierce independence who defied occupiers, stole sheep and brides, never forgave or forgot.

In 1371 the Venetians built a castle at Frangokástello to try to pacify the region. Six centuries later the solitary outpost remained humbled by the grey mass of mountains, divorced from central authority. A few miles beyond it lay Horasfakion, the region's tiny capital. During the Second World War Allied troops retreated through the battered town. Some seventeen thousand men were evacuated by the Royal Navy under constant German bombardment. Ten years later an English lady sailed her yacht into the little harbour. So touched was she by the Sfakians' support of the escaping soldiers, and by the discovery of cliff-face caves still littered with regimental buttons and badges, that she offered a gift to every child in town. The mayor thanked her and informed her that two thousand children lived in Horasfakion. In fact there were only forty.

Katrin and I twisted back and forth up the western cliffs, along a switchback road without a guard rail, its exposed edge eaten away like wafers of biscuit. The painter John Craxton, a long-time resident of Crete, had said to me, 'A modern Icarus would fly from Sfakia. He'd look out over the sea, watch the vultures rise on the thermals and jump off a cliff.' So I'd come looking for a launching point on the barren plateau.

But Sfakia was austere and unforgiving. In 1837 the British traveller Robert Pashley was assured that few locals died a natural death in Anopolis. Aradena, an adjoining hamlet which had been occupied since Classical times, was now abandoned because of a bloody vendetta. Sfakia's most notorious feud raged for fifty years and left one hundred and sixty people dead. In 1994 a shepherd from Asigonía avenged the rape of his mother by shooting all of the assailant's relatives.

Beneath massed peaks we tramped across rock-strewn fields, around ancient threshing circles, through air heady with the scent of thyme and pine. On a high crest a single shepherd with binoculars and folded umbrella watched his tan-coloured sheep. A gunshot rang out, making us duck. That spring a French hiker had lost his life in Aradena, either eaten by mad dogs or by hanging himself. The previous summer an American climber had fallen off a cliff, broken both ankles and been stranded between mountain and sea. He too would have died had a Belgian swimmer not heard his cries.

On a cliff high above the crooked shore we watched the evening sun slip behind a cloud edged in gold. The turquoise water turned conifer blue then black. Colour leached out of the olive stone and the earth glowed as if reflecting the light of the day. I lost sight of the far islands as they dissolved into sea and sky. Sfakia was a place to fly in the mind only.

Four down, two to go.

We survived the night, though more by avoiding the local *moussaka* than the vendettas. Early Tuesday morning we turned away from the sea to skirt the mountains which made the south-west coast inaccessible to cars, and drove through scattered alpine villages of mulberry trees and pitched tiled roofs.

In the hidden chestnut valleys of *Enneachora*, the Nine Villages, thick-trunked eucalyptus and plane trees shaded mossy squares flanked by an embrace of hills. Venetian mercenaries had scratched their names in the frescos of Kefali's fourteenth-century church: Huc fuit (here went) Lambardo 1530, Hic fuit Francis Leforde 1553, Draco Muazzo was here on 4 September 1588. Later Turks hacked Arabic blasphemies into the plaster. Above the town towered more current graffiti. In twelve-foot-high letters 'Drink Coca-Cola' had been daubed on the concrete foundations of a Swiss-style restaurant, with broad eaves and verandas.

The road climbed above the chestnuts, back into the sun, opening onto far-reaching views of the Mediterranean coast.

At Platanos we plunged down to Falassarna, the beach recommended by Manólis the carpenter. The broad crescent of deserted white sand at Crete's western tip was irresistible. We stripped off our clothes and swam out over the shallow, sandy shelf. The Wright brothers had first flown from a beach. And first crashed on a beach too. To our north the stubby fingers of two headlands reached from the narrow coastal plain into the Cretan Sea. But their bare hills did not protect the beach from wind, which seemed to blow all the way from Gibraltar. Katrin and I dried ourselves in the sun, in the lee of a rock, knowing that I could not fly from here either, even before learning that Falassarna had the best windsurfing on Crete.

Five down.

We turned east along the north shore. Twenty miles on, the forbidden Maleme revealed itself as the ideal spot to fly. Two runways lay at the side of the broad bay, surrounded by a flat strip of bamboo, protected from sea winds by the Rodopou and Akrotiri peninsulas. The main runway was a generous 800 metres long. The Woodhopper could take off in less than eighty feet. But like Timbaki, Maleme was a restricted military zone. Greek paratroopers in combat fatigues marched around their barracks. The smell of cordite rose from the rifle range. A pair of Corsairs swept over the airfield on a practice bombing run.

It was at Maleme that the Allies had lost the Battle of Crete and Germany – many Cretans believed – began to lose the Second World War. By late 1940 all of Europe was either conquered or neutral except for Britain and Greece. Italian troops crossed the Albanian border only to be beaten back. The Wehrmacht came to the rescue of Mussolini's humiliated army, conquering the Greek mainland in three weeks. In spring 1941 Hitler was anxious to begin his Russian campaign and so take Moscow and Leningrad before the winter. But he was persuaded to invade Crete first.

General Kurt Student, the ambitious founder of the first parachute division, wanted his elite airborne troops to see action. The island was a strategic venue for their baptism: British bombers based on Crete threatened Germany's Romanian oil wells, the Royal Navy used its harbours to defend the eastern Mediterranean and the runt of the Allied expeditionary corps had taken refuge there after the fall of the mainland.

On 20 May 1941 Germany launched a massive attack. More than 500 three-engined Junkers 52 transports, escorted by 430 bombers and 180 fighter aircraft, dropped 6,000 parachutists over Heraklion, Rethimnon, Hania and Maleme.

'Come and see the umbrellas!' called the Haniots. 'The sky is black with monstrous, terrible flying things.'

People who had never before seen an aircraft witnessed history's first airborne invasion. The 5th Cretan Division, comprised of all the island's males of military age, was stranded on the mainland. So civilians armed themselves, unearthing rusty guns which had not been handed in during the Metaxas arms requisition, attacking the paratroopers with sticks and sickles.

'You should have seen the parachutists falling,' boasted one Stamatis Borakis of Samaria. 'If you shot a hole in his parachute he'd fall to the ground like a stone.'

The airborne troops were butchered. Losses were so heavy that the invaders considered abandoning the campaign. The Royal Navy dispersed and destroyed the entire convoy of supporting German ships. The attack was defeated at every point except one, and that point was critical. The Fliegerkorps managed to capture Maleme and open the air-bridge which changed the course of the battle. Crete fell in a fierce ten-day fight. But the casualties suffered by the Luftwaffe's parachute division were heavier than the total number of Germans killed in the war to that date.

For Hitler the Pyrrhic victory marked the end of the paratroopers, whom Goering had wanted to use for the invasion of Britain. The Battle of Crete also delayed the start of Operation Barbarossa and – critically – set back the German Army's arrival outside the gates of Moscow until October 1941, by which time the early frost had begun to hinder its movements. Had the Germans reached the Soviet capital five weeks earlier they might well have captured it, and the course of the war may have been very different.

'Crete was the grave of the German parachutist,' lamented General Student. And according to local myth – if not military historians – it was the Cretans who first began to bury the Third Reich.

We drove up the low hill to see that graveyard. Hill 107 was the crucial ridge which the Germans had seized to win control of the airfield. At its crest 4,465 dead men lay beneath orderly rows of flat headstones. The bodies had been collected twenty years after the war from temporary burial sites across the island. The sombre, neat regimentation of the cemetery made it seem a very un-Cretan spot, like the shady Commonwealth War Graves at Souda Bay.

'Their deaths should be a reminder to us to keep peace between peoples,' stated the plaque beneath the single cross. I read the stones at random: Fritz Ganz 20 Mai 1941, Walter Weiss 20 Mai 1941, Oberfeldwebel Wilhelm Keller 20 Mai 1941. In an unkempt corner below the hillside was a memorial to the fifty men of RAF No. 30 and 33 Squadrons killed that same day: Eaton, Elson, Huggins and Pennystone. Nearby the bones of pashas and knights lay beneath Hania's Kastelli. A late Minoan tomb had recently been unearthed in the terraced hillside beside the German cemetery.

We turned away from the airfield.

Six down and all out.

But the memorials to former empires had given me an idea. On the outskirts of Hania was the Souda Bay NATO air base. As the Woodhopper was an American design and needed no licence in the United States I wondered if it would be allowed to fly above the American camp.

At the gate I called Captain Hoeful, the commander of the garrison. His secretary put me through to Paul Farley, the base's Public Information Officer.

'What do you need in terms of time?' he asked after I'd explained the situation. Optimistic. Enthusiastic.

'Just a couple of hours,' I said.

'Stay on the line. I'll speak to our executive officer.'

Five minutes later Farley came back to me. He said that it would be a pleasure to help me, after my flight had been approved by the Greek authorities. 'Give a call to Mr Panikitopolis over at Civil Aviation.'

'The runway isn't American territory?'

'We share it with the Greeks.' Who would be expected to ask me for a licence or two.

'I don't suppose you have an aircraft carrier in the region?' I asked Farley.

At the end of our circumnavigation Katrin and I drove into Hania, by way of the hospital. There was no sign of Polystelios or Aphrodite, so we carried on to Ariadne's house. I rang the bell for five minutes. Only then did Apostoli clamber out of bed, looking more like a ghost than a god.

'I am sorry but I did not come home until ten o'clock this

morning,' he groaned when he opened the door. 'As you can imagine it was a very, very crazy evening.'

We told him about Maleme, the only airfield which seemed suitable. He leant against the door frame trying not to fall over. 'I don't suppose that I could fly from there?' I asked him again, knowing his answer.

'It is not possible,' he said.

'And if I write a letter to the base commander requesting permission?' I persisted.

'Then I will take your letter to him,' he conceded, too weak to argue. 'When he returns from his hunting.'

'Hunting?'

'He spends his days shooting birds in the bamboo.'

I wrote my request in longhand, leaning on the bonnet of the car. The disclaimer read, 'I accept full responsibility for my actions and will not hold the Ministry of Defence responsible for any consequence of the flight.' I gave Apostoli the letter and he started to close the door.

'Aren't we going now?' I asked him.

'Going where?'

'Shopping,' I reminded him. For control wire, pulleys and the new inner tube. I was due to travel to London the next morning and wanted to be ready to fly the Woodhopper at the weekend.

'Whatever you want,' Apostoli pleaded, not wanting to disappoint. 'We do whatever you want.' He slumped down onto the doorstep. I thought for a moment that he was going to cry. 'Only please, I have not had my coffee yet.'

23. An Old Story

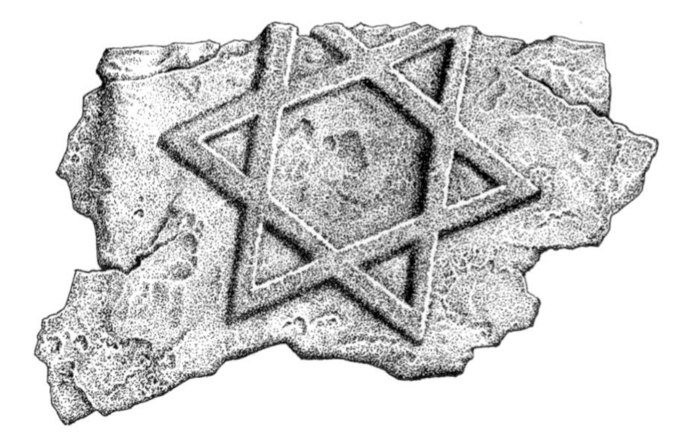

We decided to spend a last evening in Hania and have a meal. Katrin found a restaurant with a view of the sea and pigeons under the eaves.

'*Sprechen Sie Deutsche?*' asked a lanky man in chinos at the next table.

'Our English is better.'

'Thank God,' he said, pulling over his chair to join us. 'I don't feel like speaking German tonight.'

Judah, as he introduced himself, had a globular nose and a goatee beard. His skin was paper dry and the tops of his teeth had been ground straight-edge level. He was an erstwhile academic, curator and tour guide. He wore a hand-knitted *kipah* and noticed my surprise.

'You're a Jew?' I said.

'And a Cretan. My parents left before the war, shaking sand off their shoes. I returned in 1957 to save the last Jewish building on the island.'

Roxanie had mentioned Hania's Jews. She'd called them foreigners from which I'd assumed that they were more recent arrivals.

'There have been Jews on Crete since the time of Alexander the Great,' said Judah.

Almost 2,400 years.

'How big is the community?' I asked him.

'One,' he replied. 'Now only one.'

Over supper Judah told us the history.

Jews from Alexandria and Palestine had settled in Delos, Thessaloniki and Crete in the third century BC. Near Delos' ancient synagogue, he said, a stone was found with a dedication by a Jacob of Knossos. St Paul came to Crete to meet the Jews of Gortyn, which seemed at odds with the Acts of the Apostles but Judah was too engrossed in his subject to be interrupted. The second wife of Josephus, the historian who fought for Titus, was a Jewess from Kissamos. A Haniot Jew had taught Hebrew to Pope Pius IV in the sixteenth century, said Judah.

'At the peak there were up to six thousand Jews living and working on the island,' he maintained, admitting at the same time that no written record of them existed. 'Under the Venetians they prospered in trade because a Jew in Crete could trust a Jew in Venice. They also became farmers and shepherds. Kosher wine and cheese were sent up to them in the hills.'

After the meal, Judah invited us to walk with him along Kondylakis to a cul-de-sac in the old Jewish quarter. In an open, pebbled courtyard the whitewashed Etz Hayyim – tree of life – synagogue made a peaceful sanctuary. A pomegranate tree dropped its fruit on the old tombstones. Jasmine climbed newly plastered walls. Within them holy candles burnt beneath Venetian piers and gothic arches.

'I found a Turkish cannonball embedded in that wall,' said Judah, pointing across the south courtyard.

The original building had been a church, probably dedicated to Santa Katarina, which was shelled by Barbarossa, the red-bearded Barbary pirate who attacked Hania in 1538. The ruin was thought to have been given to the Jewish community a century later when the Turks occupied the island.

'It's a myth that religions suffered under the Ottomans,' Judah said, a view which was at odds with popular stories of slaughtered bishops and janissaries who, according to a French consul, lay bets on which way Christian passers-by would fall when shot. 'There was more Christian diversity than under the Byzantines or Venetians. The Orthodox church was united for the first time in a thousand years. The Ottoman Empire was quite permissive, almost like the United States. Of course the Sultan was tough on dissenters but what would the Americans do if the Hopi Indians declared themselves an independent state?'

Cats drifted out of the synagogue. An Albanian builder strolled in from the tombs, spat on the pebbles and set about tinkering with the clock, which struck the hours as Judah began to pray. He made an unusual congregation of one; opinionated, stubborn and emaciated, perhaps not from a suffering soul but rather – as we learnt later – from decades of high living in the fleshpots of the eastern Med. In Hania he had a reputation as a dissolute playboy and braggart. Yet throughout his hedonistic years he had excavated a crusader church in Jerusalem, helped to establish the Jewish Museum of Greece and written books on Judaeo-Greek cookery. He had also saved Etz Hayyim.

In 1957 Judah had found Hania to be an unwelcoming, brooding city. There were unexploded shells in the shattered *palazzi* and squatters in the synagogue. Its Venetian arches had been bricked in and shrapnel pock-marked its walls. The *mikve* was a sludge-filled pit and the courtyard served as the neighbourhood dump.

Judah began to raise money and to make repairs, first to prevent the building's collapse and then, after the 1995 earthquake, to restore it. In the early years neither the archbishop nor the city council supported his efforts and Judah earned the reputation of being a difficult man. But with donations from the Rothschilds and Lauders, and the arrival of wealthy Jewish tourists, the public attitude changed.

'What happen to the community?' I asked him when he broke off from his prayers.

'There's a lot of embarrassment because of the war.'

'Tell us,' I said.

'There are two stories and they contradict each other.'

Of course.

'29 May 1944,' he said with a single, regretful shake of his head. So late in the war. 'At four thirty in the morning the two ends of Kondylaki were blocked off. The Germans ordered everyone out of the houses – in bed clothes, crying and screaming. One person from each family was allowed back to pack a small bag. Then they were herded into trucks. It was over in half an hour.

'The soldiers ransacked the houses, throwing sheets, furniture, the lot, out of the windows. Everything from the synagogue was dumped in the courtyard of the archaeological museum. A big crowd of Greeks collected at both ends of Kondylaki. Christian Greeks. The Germans told them to take what they wanted.'

'And they did?' asked Katrin.

'Crete had gone through a terrible occupation,' said Judah. 'For months people had lived off grass and snails. It was obvious that the Jews weren't coming back.'

The 265 Hania Jews were transported to Heraklion and loaded onto the German freighter *Tanais*. They were destined for Auschwitz. One week later the unmarked ship was torpedoed by a British submarine in the Aegean Sea. There were no survivors.

Judah found a seat under the jasmine and rolled a cigarette.

'What is the second version of the story?' I asked him.

'The myth that Christians protected us,' he said.

Judah struck a match and touched its flame to the end of his cigarette. He inhaled, held the smoke in his lungs and crossed his long, lean limbs.

'I have a dear friend, Lillian Cohen, whose family was obliterated here while she was away in Paris. In the late 1950s she came back to find her property. It had been sold without her knowledge,' he went on, watching a curl of smoke rise in the air. 'The middlemen in all of this were unfortunately Jews, with Christians attached to them. Lillian started a court case which went on for ten years. Once while she was waiting for a judgement she took a trip out to Vrysses.'

'Near to where we're living,' I said.

'She and her husband stopped at a small café to have coffee and an old Greek woman came out and started talking to her. She thought that Lillian was a foreigner and invited them into the house behind the café. And there right in Lillian's face was her mother's Sèvres china, marked with the initials "LC". Lisa Cohen.'

He dropped the half-smoked cigarette onto the pebbles and ground it into the stones.

'The old Haniots are either ashamed that they did nothing to help us,' said Judah, 'or they took part in the ransacking.'

He picked up the butt and began to rip it apart with thin, hardened fingers.

'Plus we now have the incomers: the raw, rough, ignorant

shepherds who moved into the vacuum left by the Turks and Jews. They have no history here. They bought our vacant properties for nothing. They sell tourists trinkets from Hong Kong and kill the goose which laid the golden egg.'

It was understandable that his bitterness had been turned against the enchanting Venetian town. His anger, like his synagogue, was part of its fabric. Similar strands of enmity wove back through Crete's history: the exiled Turk despising the conquering Christians, the defeated Byzantines hating the Arabs, the Minoans disdaining the uncultured Dorians. Defiance was central to the island's identity. After the Battle of Crete a German paratrooper reported that the Cretans were barbaric fighters. 'They didn't give us time to land and take up position,' he despaired, haunted by the memory of his helpless comrades shot as they drifted to earth. 'We were told that the Cretans loved visitors.'

When in 1999 Judah finished the restoration the Chief Rabbi of Salonika asked him about his plans for the building. 'Do you want it to be a museum?' he asked.

'No,' replied Judah.

'Then a memorial to the deceased?'

'I want it to be a synagogue, like it was.'

'Then you will have to pray here every day,' the old rabbi told him. So Judah became Hania's rabbi, or 'almost rabbi' for, as he admitted, the appointment wasn't wholly kosher.

'When this building falls down,' he said, lifting his arms towards the filigree Star of David which had once served as a chicken coop, 'it will be the end of all memory of the Cretan Jews.'

There had been seven synagogues on the island. All the others had been destroyed, including the oldest which lay buried under the Xenia Hotel in Heraklion.

'Crete was once proud of its racial and religious diversity,' he concluded, drifting through the centuries like a feather in the breeze. 'The *muezzin* called the faithful to prayer, black Sudanese slaves fished with trained cormorants, Athenian women visited to buy the latest fashions. It was an international city. Now they hold motorcycle exhibitions in the mosque. One people. One language. One tradition. The Greeks haven't begun to realize what they've lost.'

We left him in prayer, leading his community of one, honouring the dead. He transcended loss and made meaning out of that which remained.

Beneath a full, yellow moon Katrin and I circled the harbour back to the car. In the side streets the mopeds were parked three deep. Their young riders crowded the waterside cafés which exhaled warm, smoky breath into our faces as we passed. Dark-eyed girls in knee-high boots and black micro-skirts queued to enter Club Barbarossa, the pirate now inside their city's walls. Their dates drank frappés, sat in large groups and sent text messages to one another. In one bar I caught sight of Apostoli, with a long-legged woman on each arm, laughing and – probably – telling stories.

24. On a Wing and a Prayer

Yióryio hammered on the shutters, rousing me from a dream. I'd been flying bump-and-circuits above Anissari, touching down on the church roof to swallow glasses of *tsikoudiá* and motion sickness pills. My engine's knocking became his hammering and clipped my wings. I whacked the clock off the bedside stack of How to Fly manuals. I was late. We'd arrived home from Hania after midnight. My plane left for London in under two hours.

'Telephone,' called Yióryio through the door, his voice subdued. He must have overslept too. 'From England.'

While Katrin shoved a change of clothes into a carrier bag I belted down to the *kafeneion*, blinking in the sunlight. Loony Ulysses was walking backwards along the street. The receiver lay on the counter.

'Hello?'

Westlake-Tom's secretary at Meggitt was on the line.

'Sylvie, we're just leaving for the airport,' I said. 'I'll be with you by late afternoon.'

'I'm sorry but it's not good news,' said Sylvie, stopping me in my tracks. 'We've spoken to our lawyers. We can't loan the engine to you.'

'But I signed the disclaimer,' I reasoned. 'I'm willing to take responsibility.' For my life.

'I'm sorry,' she repeated, 'but because the engine isn't licensed for manned flight your personal indemnity does not abrogate us from our obligations as manufacturer and designer.'

She was quoting the lawyer's letter.

'If there was an accident you – or your relatives – could come back at us and claim that we were at fault. I'm sorry not to have told you sooner.'

I put down the receiver and stared out of the window. Katrin pulled up in the car and sounded the horn. It shattered my reverie and the hush of the *kafeneion*, silent even though every table was occupied. The only sound was the hiss of the gas flame beneath the coffee *bríki*. The drinkers were dressed in black, or more black than usual.

'The dogs were howling in the night,' said Yióryio. I didn't understand him. 'Aphrodite is dead.'

The operation had gone wrong. The villagers, who knew their neighbours' smallest intimacy, including the secret niche behind the church where Iánnis hid his pornography, had told me nothing about her deteriorating condition. They had wanted to spare me, or maybe still felt me too bad tempered to take in another's grief.

The room was lit by low black candles set alongside the coffin and stretching away to a funereal infinity. Or as far as the privy. The floor was a void, veiled in dirt. Dark cloths covered the radio and television. The bench covers were stained and soiled. The mirror faced the wall.

Polystelios sat alone in the gloom in an inky suit, picking at the frayed cuff of his blued shirt. The smoky candlelight caught only his face, a plaster white mask adrift in space. He had built the rough box himself, dismantling her family's armoire and part of the chicken coop, lining it with the moth-eaten rags from her trousseau. He'd made a pig's ear of the job. The edges weren't square and he'd run out of paint but had refused all offers of help. He'd washed her ample body in wine and dressed it by himself. In death at last he would have her to himself.

'Won't you open the window?' I asked him.

The room was stifling, its shutters bolted against the light. An unpleasant smell tainted the air.

Polystelios had intended making the coffin wide enough to take her bulk but in his sorrow he had reversed the measurements. Aphrodite lay on her side, neither angelic nor at peace, pressed into the narrow crate with her shoulders around her ears and breasts kneaded under her chin. She looked like one of her fatty offal dishes dressed up in cotton grave clothes. I did not want to kiss her.

'Before she died she called their names,' Polystelios said, pulling himself onto his feet, not looking at me. 'All their names.'

Her lovers? I wondered. Her parents? Or the children she never bore?

'I felt her words fly through the air and come into the rocks. Now they're all around me.'

I moved to stand beside him, not comprehending him, and laid my hand on his shoulder as Greek men do.

'She called your name too.'

'I didn't think she knew my name,' I said.

'*Pilóte*. She called you, "*Pilóte*". She wanted to see you fly and to fly with you.' He hobbled closer to the coffin. 'But for that you would have needed a bloody big engine.'

On Wednesday morning six strong men carried Aphrodite on their shoulders out of the filthy house, just as she must have dreamt countless times. The coffin was open to the air and wild flowers fell in its wake. The scent of crushed blooms rose with the keening of a group of mourners, dressed in drab dirge-clothes.

'I don't recognize any of them,' Katrin whispered to Sophia.

'Unwept, unburied,' she replied. 'They're hired professionals.'

Greeks have long believed that the departing spirit was a sure and unerring messenger to another world. Funerals presented an opportunity for the living to communicate with the dead. But Polystelios claimed to have no relatives and his only friends were alive and drinking at the *kafeneion*. So Yióryio raised the money to hire the mournful procession. The village would not be denied a funeral.

'Her light is stifled by death,' sobbed the wailers, clawing at their hair and stroking the coffin. 'Lord have mercy upon us.'

We followed them, every corner of Anissari echoing with the wildness of their grief. It rang through the olives, bawled along the shore, moaned in the high mountain gorges. In no time it moved the local women to tears. Then loony Ulysses began to mewl, gripping the seams of his trousers and yanking them above his black shoes.

'The dead stay dead,' wept the professionals. 'Have mercy upon us.'

On the other hand the village men – each of whom had

claimed to have been Aphrodite's lover – talked among themselves about sex.

'Why do you have no children?' Little Iánnis asked me, though everyone understood at whom the question was aimed. His voice chaffed with disdain. 'Maybe – like Polystelios – you eat only vegetables?'

'A man who eats salad cannot look after his wife,' said Kóstas as the death-doused cortège whined by the garage.

'You need meat for your own meat,' taunted Iánnis.

'I know,' Polystelios suddenly interrupted, not taking his eyes off the cockeyed coffin. 'I know you rolled my wife. All of you. She told me, over and over.'

Tears dropped on Aphrodite's dark head. No doves cooed in the orchard. For a moment the men walked in silence, except for Ulysses whimpering gibberish. They were not used to Polystelios taking a stand.

'You spores of the Turk,' Polystelios spat at them, helpless in his misery, his heart beating faster than a rabbit chased by dogs.

'A good story makes a better truth,' said Yióryio, in an attempt to leach his anger.

'Of all you can say nothing to me,' Polystelios hissed back at him. 'You screwed her last, on the street after the *panegyri*.'

'Don't believe those stories.'

'It's all the same to me,' he said. 'I know the truth.'

In fact none of them had slept with Aphrodite. Ever. Not even the priest. That at least was the tale which would now be told in the village.

The bearers bent under the weight and the boughs of a jacaranda tree to heave Aphrodite feet first into the cemetery church.

'May her memory be eternal,' cried the women.

In Crete bedheads faced the east. All churches too were aligned to the rising sun. A lifetime's orientation changed at death. The departed were buried with their feet to the east and their head following the setting sun.

Papá Nikos led the prayers, his eyes wet, distressed as I'd not seen him before, and the mourners, professional and amateur alike, were shaken by the ceremony. He dispatched Aphrodite's soul to God and her body to the earth as the congregation worked themselves into hysteria.

'Lord, have mercy upon us.'

Only three slabs had been lifted off the top of the tomb so the coffin had to be slid in by the head. But because of its home-made dimensions the angle was too acute. Aphrodite's flabby white arm swung out of the open box hitting Socrates in the face.

'Bring me a shovel,' he called above the weeping, intent on levering the offending appendage back into place.

But Aphrodite's parting wave proved to be too much for Sophia. With a woeful howl she fainted across both the coffin and the grave.

'Lift her legs! Get blood to her head,' shouted the bearers, trying not to lose their grip.

Manólis and Iánnis lifted Sophia by her feet and her dress fell down. Yióryio, who had been looking for a shovel behind the iconostasis, saw the men, Sophia and her blue bloomers and yelled, 'Let her alone. She's my wife.' They dropped her on the ground and, as she staggered to her feet, Yióryio whacked her across the rump. 'Get back to the house, woman,' he ordered and she went like a dog.

Until recent years it had been the custom for a Cretan widow to don a mantilla, the black peaked headscarf which would be worn for the rest of her life. In the depths of her mourning she might pull out her hair or a tooth. She would wash her husband's grave every morning. She might never again eat meat.

In contrast a widower would probably remarry. But that seemed unlikely for Polystelios – hopeless Stelios-of-Many-Talents – as he surrendered himself to grief. When his fleshy wife was levered into the grave his impish plaster face dissolved in tears of loss.

A hired mourner lay down and hammered the stone with her fists, then tried to crawl into the tomb, until she was pulled away by her partner. Polystelios stood silent, stifled by death, as the three slabs were dropped into place.

Fifteen minutes later Yióryio had convinced Polystelios that the stories of infidelity had been told in jest. In a seamless, instinctive transition from sorrow to acceptance he poured a glass of wine for his friend then embraced him. Leftéri played with his new toy helicopter under the *kafeneion* tables. Across the square Socrates started his chain saw to prune olive trees. The hired weepers stood in the sun eating cake.

'Let the earth rest lightly upon her,' said one of them to Polystelios, shaking his hand.

'As she rested heavily on me,' whispered Yióryio.

'We *have* to finish the Woodhopper,' insisted Apostoli later in the *kafeneion*.

'But there's no engine,' I said with resignation, 'nor is there any hope of one.'

'Never say never. There is always more hope.'

And more broken promises.

'How did you get on with Maleme?' I asked. Yióryio listened from behind the bar.

'I'm sorry but you have no permission,' Apostoli admitted, apportioning blame with 'because you wrote that you will accept responsibility.'

'Which is true,' I said.

'That made them think that something may go wrong.'

All the *kafeneion* was now listening, whether they understood English or not.

'So we have neither an engine nor an airfield,' I repeated. In the aftermath of the funeral I felt defeated.

'The big companies won't help us,' he replied, as if he had always known that Meggitt would renege on their promise and wondered why I hadn't too. 'They already have their own aeroplanes. So we must look for help from little people.'

'Who, Apostoli?'

'We are born with brains so we can think. Let me think. I have a friend in Thessaloniki who knows a hang-glider shop.'

'And?'

'He may sell engines. Also, I know a man who is building his own aviation museum. His name is Christodoulakis and he can find everything.'

'Manólis Christodoulakis?' asked Yióryio. 'He is my uncle. His grandfather born in next village.'

Apostoli lifted his shades and looked me in the eye. 'This is good. I will call him.' He smiled. 'You ate the donkey so don't leave its tail.'

'Pardon?'

'You finished the body. Now we find an engine, even though it is the most difficult part to eat.' The drinkers nodded at his wisdom, which encouraged him to add, 'In the old times when Greeks had wars they used swords and shields. Mothers told their sons, "*I tan i epi tas.*" Be with the shield or be on it.'

'I am with the shield,' I assured him, 'eating a donkey's tail.'

I did want to fly, but now it was for the living as much as for the dead.

'I'm going for a swim,' I said.

'After flying practice,' replied Apostoli.

★

He had brought his computer. We set the monitor on a workbench beneath the Woodhopper's nose and taped the control stick into the cockpit. The loudspeaker was suspended from the engine bracket. I put on my bicycle helmet and decorator's goggles, slipped between the catscradle of wires and settled into my seat. Apostoli rebooted the computer and selected Hania airport. Onto the screen flickered the arrivals building and control tower. The sound of an idling engine ticked above my head. I was at the controls of my own aeroplane, preparing to fly, inside a Cretan garage.

'I have written out your instructions,' said Apostoli, attaching a sheet of paper to the A-frame. It was headed 'How to Fly'.

'One,' he read aloud, 'find the middle of the runway.'

I taxied forward, swinging back and forth across the motorway-wide apron like a drunken beetle.

'Keeping to the centre is better,' said Apostoli.

'I'm doing my best just to stay on the tarmac.'

'Two: check the elevator and rudder. Three: check the throttle.'

'Do you mean increase throttle?'

'To 100 per cent. Softly.'

'Here we go.'

The runway's white lines began to slip away under the screen and my feet. I began to pick up speed.

'Stick forward,' said Apostoli.

Ten knots. Twenty knots.

'Ease back . . .' I said out loud, '. . . and up she goes.'

The ground fell away. I was airborne, already drifting off course but flying better than last time, watching my on-screen instruments.

'Keep your nose down. You need less throttle. Seventy-five per cent,' said Apostoli.

I rose above sun-burnt fields and lush valleys of vines. Below me a computer-generated mule trotted away from the noise of my engine. The sun was setting over Akrotiri. It glittered off the silver sea until I was blinded by the glare.

'Keep your nose down. Less throttle,' Apostoli repeated.

'Hopefully I won't crash.'

'You must not crash.'

I told him that the world's first two mail aeroplanes – one French and one British – flew between London and Paris. They were the only aircraft in the sky yet one day they collided.

'I don't know if it's true or not,' I admitted.

'It doesn't matter; it's a good story. But don't lose your concentration. You are still learning.'

'I'll turn now,' I said.

A gentle turn to the right over the Cretan Sea. About five degrees. Flying reminded me of mathematics, of equations and angles. In my little machine I stretched straight lines across the curved surface of a disordered world. The variables were infinite.

'There's Hania,' said Apostoli.

In ten minutes I'd begun to fly straight – ish. But Apostoli introduced wind into the programme and every updraft knocked me off course, as did oversteering and lapses of concentration.

'You're pushing on the right rudder,' he said. 'Let up on the controls. The aircraft will tell you what to do.'

'The nose is pointing that direction,' I told him, motioning to the right, 'but we seem to be flying this other way.'

'Good,' he said.

I completed the circuit, drawing a rectangle in space, then on the base leg throttled back to 4,000 rpm. Ahead I saw the four-kilometre runway. The lights were on.

'Nine,' read Apostoli from his instructions, 'Release some

throttle but keep your altitude. Move the stick in small movements.'

The secret was not to be timid, to be light and fluid with the controls, to keep eyes ahead and not to exceed 35 mph.

'For now your approach is good. Turn left softly. You are at about two hundred metres. Release some throttle.'

I tried to imagine myself rolling down a hill, drawing a line from the cockpit to the end of the tarmac. I made the runway come to me, not me to it, monitoring the angle of descent by its relative position on the screen. When the runway appeared to move up the screen then my descent was too steep. I applied a breath of power.

'Push the nose down and follow the line,' said Apostoli.

The runway rushed towards me, growing in the screen. There was still a kilometre to run.

'I think I'm too low,' I said.

'You are very low,' he answered.

As I crossed the perimeter fence at fifty feet I couldn't stop myself from pulling up hard on the stick. My instinct was to fly away from the ground.

'Do that and you die horribly.'

I dropped the nose a fraction. Stabbed on full power. The Woodhopper would stall at more than fourteen degrees incline, without enough height for me to recover.

'Go up,' said Apostoli. 'Up around again.'

Little Iánnis stuck his head around the door, saw the flickering monitor and asked, 'What are you doing?'

'Flying,' answered Apostoli, as if it wasn't obvious.

'Where's the engine?' he asked.

'We don't need one,' I said.

'So now you can go home,' he replied. 'Goodbye.'

I turned back into the circuit, setting a course, allowing for drift. I followed the coast, banked above Maleme and returned to the north of Agia Theodori. I tried to keep the computer-generated aircraft in equilibrium, lift balancing weight, thrust balancing drag. The virtual wind was gusting to fifteen knots. There were pockets of rising air. I was all over the sky. Or the garage.

I made my final turn above the airfield beacon, throttling back again to 4,000 rpm. Once more the descent felt steep but I made myself stick to it. I wanted to flare out at two feet of altitude. I concentrated on the end of the runway, choosing my point, drawing my line through the sky.

'You're doing it,' said Apostoli, surprised, watching the runway come straight towards us.

I crossed the perimeter fence at less than forty feet.

'Back pressure on the stick.'

The nose lifted. I levelled off at ten feet. I was flying above the tarmac, dropping foot by foot.

'More back stick,' instructed Apostoli.

The Woodhopper's nose lifted higher, but without extra power the machine began to stall. It slipped down onto the ground. I was no longer a bird but an earthbound machine, motoring along the tarmac at 22 mph.

'Full power,' said Apostoli.

'Am I going up again?'

'Of course.'

Six thousand rpm. A touch of back pressure on the stick. The aircraft swept – rather than jerked – into flight. I climbed to 800 feet and turned back into the circuit.

'Now you are flying the plane,' brightened Apostoli, 'it is no longer flying you.'

I practised a dozen circuits, watching my rate of climb, finding level flight by relating the nose – or television – to the horizon – or whitewashed wall. I increased speed, climbed, then levelled off, trying to make flying an instinct. I landed, blew a tyre and took off again, learning to ease on the speed, to push the stick forward, to keep nose down and tail up.

I stole a look at the Cretan Sea and the Lefka Ori. A bird swept towards me. At first I thought that it was part of the software, an added feature like the mule on the simulation programme, until another bird flew past the edge of my vision.

'Don't lose your concentration,' said Apostoli.

'They're swallows.'

'Of course,' he replied. 'It's May.'

The swallows had returned to Crete, swooping and scudding along the road and into the garage, around the wires, looking for nesting sites. A year before I had sat with my mother in the green English bedroom, watching the swallows and house martins nest under the eaves.

'Keep flying,' Apostoli repeated to me. 'You are flying your aeroplane.'

And I was, among the birds in the white Cretan garage, laughing.

Later we switched off the monitor and loaded the computer back into his car. I felt exhilarated and nauseous from too many circuits and on-screen spins. Apostoli was pleased.

'Now you need a weekend's holiday,' he instructed me. 'I will find the engine for us. No problem.'

I wanted to believe him.

'I drop you off at the beach,' he offered.

'I'll walk,' I said to him. Wanting to feel my feet on solid ground. 'And think about life.'

'What is there to think about?' said my bat-eared Greek god. 'The life is good.'

'You must fly like Hermes,' Yióryio told me, picking up his sheep shears, 'like first *pilótos*.'

A dozen village men dallied in the open, cadging cigarettes, their voices warming in the morning sun. Beneath us spread the flat roofs of Anissari. Behind us their sheep were penned together into a cleft of hill.

Kóstas wore camouflaged hunting fatigues. Papoos shared a joke with two shepherds from Vrysses. Socrates took the first sheep from his son, carrying it by two legs like an ungainly sack. He laid it on its back and collected its hooves together, knotting them with a length of rounded cord. Then Yióryio stepped forward to shear the wool by hand, pushing the blades against the skin. Above them on a stone perched Ulysses, his broken radio pressed to his ear. These were the people who had cared for me, helped me, enabled me to fly away. The community, where everyone knew everyone else, had become my *paréa*, a kind of temporary family.

'The first *pilótos* built his aeroplane under those trees,' shouted Yióryio as he worked.

According to him the aviator had arrived on horseback in the village, found its broad fertile bowl and distant hills to his liking and decided to stay.

'And where did he fly it?' I asked him.

'Down there outside the *kafeneion*,' insisted Papoos, throwing away his cigarette to help with the next animal. 'There were not so many olives then.'

It was possible, of course, as was ballooning over Everest and winning the lottery.

'He cut bamboo in river and buy cloth from fishermen.'

'And the timber?'

'From my father,' called Manólis, skipping a generation or two. I watched him shear along the line of wool, cutting fast, lopping off the matted fleeces and exposing mottled patterns of bare skin: flesh-pink, speckled brown, patches of blood-flecked white.

'The *pilótos* also had a lathe,' said Socrates, bringing forward another animal, the sweat glistening on his pate. He didn't approve of my store-bought ironmongery. 'And he spoke better Greek than you.'

'You see,' said Yióryio, 'you are not the first.'

The villagers warmed to their work and story, weaving the aviator into local history with strands of metal wire and impudence. He had brought the first bicycle to Apokoronas. He had rented a house from Christians and fought the Muslims. He had helped to shear their grandfathers' sheep. He had married – as Roxanie had also suggested – 'a beautiful Cretan girl' named Persephone who lived half the year overseas. Yióryio even swore that he had seen a photograph of the happy couple, standing in front of the aircraft, although, of course, he couldn't lay his hands on it at the moment. Each tale began with the now familiar refrain, *apístefto kai ómos alithinó*, the unbelievable twisting around a deeper truth.

In the same tapestry they had stitched Zeus the immortal dying near Knossos, Poseidon fusing into Agios Nikolaos and Alexander the Great's sister becoming a mermaid. When first in Anissari I had tried to unpick fact from fiction, to distinguish the timely ('the Minoans arrived from Anatolia in 2600 BC') from the timeless ('Zeus ravished Europa beneath a plane tree in Gortyn'). But it was impossible, for the Greeks themselves honoured their legends above factual history. Even the Greek Air Force was not immune to flights of fancy, calling their pilot-training academy the School of Icarus. Malleable myths linked the present to the past, sustaining a continuity across generations, creating the illusion of eternity.

'When did the first aeroplane fly in Crete?' I had asked Yióryio during my first week in the village.

'Before my grandfather's time.'

'And when did Icarus fly?'

'Before my grandfather's time.'

On that first blustery morning in the ruins the villagers had understood my need and, anchored in their own culture, had fed me a myth in order to lead me to find redemption. Their myths taught that there was order in the universe, that life had meaning, that we weren't airy spirits but creatures of narrative, of earth and time. They had shown me that I was not alone.

'I will really fly the Woodhopper,' I assured Yióryio as he sheared the last sheep, as men had done for countless generations.

'And I will tell you another story,' he said.

25. Summer Time

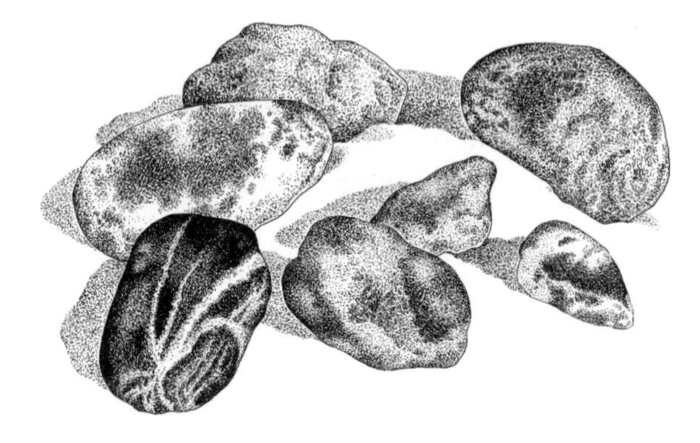

Swathes of rose-pink oleanders clustered up sage-green gullies. Hollyhocks and wild blue chicory rustled in the sizzling breeze. Every morning we woke to the hacking of hoes. Each night we slept beneath a mantle of sawing cicadas. Five months had passed since our arrival in Crete. There was no longer coolness in the air, neither in the early morning nor at night. Polystelios sowed peas, cut cucumbers and put his sheep on the high plateau. Farmers harvested melons and tender little squashes. Instead of carrying oranges, pick-ups passed the front door with stacks of deck chairs. Charter flights circled above Hania and tour buses queued along the road to the Samaria gorge.

One coach paused by the village square and disgorged its passengers. A woman noticed the white wing tip, called to her husband and when we looked up from our work twenty curious English faces stared in the garage door.

'It's one of the most beautiful things I've seen,' said an accountant from Luton.

Katrin and I tightened the stabilizing wires, which was like trying to tune a rubber piano. Each wire affected its neighbour and adjusting one meant changing three others. There was little else we could do until Apostoli brought the engine. We sat in the sun and let control slip out of our hands.

Next morning the heat had the fierceness of a blaze. I opened the shutters onto sweltering summer smells: baked earth, stinking fennel, the disorderly upthrust of growth in the mulberry trees.

'You call this hot?' said Yióryio, hosing down his patio and his daughters' dancing feet. 'Wait until July.'

The *kafeneion* tables had been moved across the street under the shade of the olives. Leftéri cycled around them chanting to his birds, '*Hélios! Hélios!*' Sun. Sun.

'The sea,' Katrin said to me.

Little Iánnis had told us about a hidden cove tucked into the waist of mountains. We drove out in search of it.

Beyond Paleloni an unmarked road wound down through a scattering of olive trees. We followed its uncertain descent, traversing the hill in lazy switchbacks, between hummocks of orange and rust bushes which masked the rocky terrain. There were no buildings along the road, apart from a Hellenic Navy listening station. I saw no farmers or sheep. At every turn we expected to meet a dead end.

Then the track fell away and we dropped towards a brilliant turquoise sea. The broad bay was a ragged semi-circle, rounded like the head of an octopus. Green tamarisk trees shaded a deserted concrete jetty. Across the inlet a sheer rockface lifted up to the familiar mastodon back of Drapanokefala.

Katrin and I picked our way over the rocks, around miniature sea urchins and launched ourselves into the clear waters, skimming the surface in a shallow dive. The sea was silky warm on the surface and above the white sandy floor, but cool in its depths. Needle-nosed fish darted between the rocks and reeds. I remembered my first swim in Crete, on the edge of tears but unable to cry, weary, colourless and flat.

I spotted a scuffed white kayak on the shore, tucked in behind a stone outcrop, and swam back for it. The boat was sound and I paddled it out to Katrin. She slipped onto the bow like a golden seal and we sculled above the reef, following the rim of the shore. The mouth of a cavern opened before us and we glided into its spangled interior, into the rocky body of the earth. Great boulders were clasped in the roof above our heads. Waves slapped and gurgled against the glittering stones.

I paddled back into the daylight. Now the opposite rockface took on the look of the train of a velvet robe, tumbling off the mastodon's back and into the sea. As we drifted towards it I noticed the sharp angle of man-made objects at its hem. I stared and made out bamboo railings then a bridge, deck and fluttering Greek flag rising above a rocky stern. I blinked. It was a stone ship, and next to it was moored a wooden rowing boat. Above them, carved into the side of the mountain, was the inscription, Paradise of Vaggelis.

The bay wasn't deserted. On the far shore appeared a slender, bearded figure with a dog. He wore only a red Speedo swimsuit.

I shouted, 'Are you Vaggelis?'

'*Ela!*' he called, waving us forward. Come.

There seemed to be nowhere to land on the sea-nibbled edge. He directed us around a ridge of submerged rocks. As we drew near the dog darted back and forth along the shore then threw itself into the water to reach us.

'Welcome to my *Parádisos*,' Vaggelis said.

We lifted the boat onto the tilting rocks. Katrin didn't have shoes so Vaggelis paced back to find a spare pair. We didn't want to intrude on his solitude – he hadn't moved to Kefalas to meet strangers – but he seemed to be a most accommodating recluse.

'I will make this a path one day,' Vaggelis said as we followed him across the boulders to the stone ship.

At first the ship appeared to be an impractical, casual shelter. Its roofed deck was built against the cliff. A table and doorless refrigerator hung with frying pans created its galley.

'Does it sail far?' I asked him.

'I stand on the bridge and go everywhere,' he replied.

In contrast with his angular body, Vaggelis' facial features were soft: rounded lips, curved chin, rock-pool eyes blue like the sea. His voice was gentle too. He seemed to prefer to live apart from the world, sailing out into it only at the flotsam wheel of his stone ship.

'Without worries about the wind or an engine,' I said.

'There is more,' he said.

We climbed a split-log ladder above the aft deck, thirty feet up the cliff face to a wooden parapet, built within a curled hand crag under a caper tree.

'The goats are a problem,' he said. 'They walk across the top of the cliff and knock rocks on my head. And when I'm out fishing they shit in my kitchen.'

We stepped over a threshold mosaic of innumerable polished pebbles into a cave. His home was the size of a two-car garage, domed and arranged into rooms by the natural shelves of rock. Carved stone columns supported the upper lip of its catfish-mouth door, wide and not very high. Open to the light, it was dry and protected him from the elements. Stalactites clung to the slope of the ceiling. The warren was orderly and scrubbed clean.

'I am the first man to live in the bay,' he told us. 'I built my first house in a cave across the water then moved here to *Parádisos* when it became too busy.'

'Too busy?' asked Katrin.

'People swim there at the weekend,' he said.

Vaggelis led us to a low stone table with hand-made wooden stools. Sea shells were displayed on square shale outcrops. A spade-shaped broom hung on the wall. A wire hook hooped through a crevice held a beachcombed basket. In the alcoves were oil lamps and favourite stones. Lots of stones.

'I bring the pebbles up from the beach. The rocks I carry down from the mountain.'

'By hand?'

There was no electricity or machinery in his paradise.

'I hold the stones like my children.'

Vaggelis pulled a round marble stopper from the neck of a clay jug and poured *raki* into ceramic eggcups. He opened a screw-top jar of peanuts and the dog, hearing the cracking of shells, came begging at our feet.

'It took me forty years to see my love,' he said and for a moment I didn't understand.

Vaggelis was born in nearby Paleloni and left school at the age of twelve. He moved to Athens, drawn by the promise of prestige and security. He worked in a chrome-plating factory for twenty-seven years.

'Every day I hated it. Every day I thought of Crete. I turned forty and my life was nothing. So when the factory closed I came here to think. Because I had always loved this place. Slowly I understood where to make my home.'

The idea had evolved during his daily walks.

'I started to know the rocks and to love them. I saw how to combine them in my own way. Then *Parádisos* called to me. It called to me across the water.'

'You live here year round?' I asked.

'All but November December when I work in the village picking olives. That gives me enough money to live. Almost enough.'

On one wall hung a slate plaque. Carved into it were the words, 'I love the place I was born. I love the rocks that make my feet bleed as I step on them.'

Behind us the cave dipped away into shrinking openings, into darkness.

'In the afternoon when it's very bright I go back there,' he said. 'Will you see?'

We descended through a hole in the rockface and into a space between the stones, crawling head-first and monkey-like through a tiny aperture. A miniature door opened into a broom-closet space, into which we unfolded ourselves. It was impossible to stand in either of the two subterranean hollows. The first room was a kind of parlour, large enough for three people to squeeze together on its stone ledge. At its centre was a table, made from an upturned marble roller once used for pressing soil on earth roofs, and decorated with dried flowers and photographs of a woman. The second room was a narrow-crawl-space bedroom with mattress, two neatly folded red blankets and a small, tidy collection of books. The ceiling was eroded, bulbous and heavy. It was a space that seemed to be suspended by imagination alone, that shouldn't exist.

'I like the quiet,' he said. There were icons on the wall. 'I like to be alone to read, to play my guitar. And to work; that is what I love. The day that I stop working with stones is the day that I die.'

I asked him if he worked to be remembered. 'What will happen when you die?'

'I must have ten more years to finish this work,' he said.

He had already spent a decade burrowing into the rock.

'After I am dead maybe the work will stay. Maybe someone will destroy it. I don't mind. I do it for now because it helps me to see.'

As he spoke he touched the stones and his chest.

'People ask me why I make this work for no money. And what I do here all day. They don't understand. They are *maláca*. Wankers. The payment for me is in the making of something.'

A body-like opening connected the two rooms, shaped from the natural column of the cave. Vaggelis slithered into his bedroom to find more *raki* and coffee. In his own time he boiled water, carried in casks from the village, and made us drinks.

'I set myself apart,' he said, 'to get to the heart of things.'

The cramped, cool hollow did not feel claustrophobic, perhaps because of the wonder of its hewn interior, of the inhospitable space reworked to a human scale, of the simplest of creature comforts. There was a rightness to the place, even though its creator was quite mad.

'Are you on holiday?' he asked us in his unhurried, sleepy voice.

We explained about the aeroplane. Vaggelis listened, asked if I'd ever flown before and then – quite suddenly – he and I laughed together, the hermit and the *pilótos* recognizing the other's compulsion.

'Everyone in the village wants him to achieve his vision,' Katrin told him.

'It is a dream,' said Vaggelis with a smile.

It had always felt like a necessity.

'You will fly from Anissari?' he asked.

I shook my head and bumped it on a rock. Minute grains of sand dropped onto my lap. 'I don't know where to fly,' I said.

'There is only one possibility,' he said. 'You need to go away from people and roads. Go to where Zeus was born. To the Nida plateau.'

The highest place on Crete.

In the subterranean burrow we talked for an hour of high places until the stony cold began to penetrate our bones. We crawled back to the main cave and into the light. At the catfish door Vaggelis asked us to sign his guest book. I'd never met a hermit before, much less one with a guest book. I leafed through its few pages. We were his first visitors in six months.

As he helped me to push the kayak away from the rocky shore, his dog plunged into the water to follow us. Katrin and I paddled away, enlivened by the prospect of a heavenly place to fly.

26. Yes of Course Maybe Why Not?

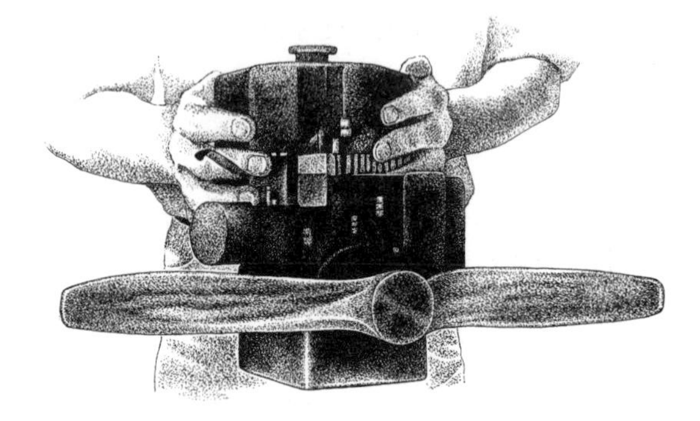

'We buy you perfect motor,' said Yióryio. He stood at the garage door with Kóstas and Socrates. In their arms was a shiny new Minirelli engine. 'You must fly,' he said. 'So this we do.'

'Thank you,' I said.

'Do not say thank you. This is the Cretan way.'

Yióryio did not want any guest of his village to be displeased. Nor his friends at Cretan Television. So he had asked Socrates to call his cousin in Sfakia. The cousin's neighbour worked for an agricultural supply firm. He found the engine and Yióryio took up a collection in the *kafeneion* to buy it. Kóstas had driven over in the police car to collect it. It was an unexpected, generous Greek solution.

'And after you fly,' Yióryio added, 'we use it to run wood saw.'

Unfortunately the engine was small. Very small. And in the afternoon Apostoli seemed more shocked than pleased.

'Will it work?' I asked him.

'Yes of course maybe why not?' he replied, turning the device over in his hands. 'What power has it?'

'I forgot to ask.'

'My mother will test it on the bench. With the propeller.'

Somewhere Ariadne had found a spare propeller. Which was a relief as Apostoli's inquiries had come to nothing. The hang glider shop had closed and Manólis Christodoulakis, the local man who was building an aviation museum, had become a monk.

Word of Apostoli's acceptance of the Minirelli spread across the village like water into cracks. Kóstas' brother, who drove the district's ambulance, stopped by with his girlfriend. The mayor called in to discuss plans for the party.

'So you fly this Saturday?' he asked. 'It is a big day for us.'

I was in the garage early Tuesday morning correcting a warp in the subframe, cutting a new length of aluminium tubing to run between the seat and the vertical strut. The Woodhopper's shape was changing with the rising temperature, expanding in the day and contracting at night, its wires loose then taut like my emotions. I began replacing the temporary nuts with self-locking nuts so everything would be as ready as possible. I had to fly soon, or at least before the aeroplane fell apart.

Apostoli didn't call all day.

On Wednesday a heat haze veiled the hills, stupefying the villagers and browning the grass. Nothing moved in Anissari, including Apostoli. I needed his help to make the control system. The summer winds might begin at any time. In the afternoon I sent him a text message. 'Are you coming today? Tomorrow?'

No response.

'What's news?' shouted Yióryio as I walked by his door that evening. 'Is ready?'

'Apostoli is late,' I replied.

On Thursday I left him three voice-mail messages.

A national strike was called for Friday in protest to changes in the state pension plan. It had been timed to create a long weekend, and thereby win maximum support for the action. The retired men marked the day by getting drunk earlier than usual. They were celebrating because the new restrictions wouldn't affect them, as they wouldn't come into force for thirty years.

At noon Apostoli replied to my messages. 'I cannot come today,' he texted me. 'No gas. No gas stations.'

On Saturday I learnt that Ariadne was having difficulty attaching the prop to the engine. 'Because the machinist made a hole in the wrong place.'

On Sunday when I called, Apostoli picked up the receiver. 'Don't worry,' he said, 'but my mother has broken the propeller.'

On Monday Apostoli arrived without the control wire and in a foul mood. 'I do not like the engine now,' he complained. 'It is from a wood chopper.' A chain saw.

'But we're building a Woodhopper,' I said.

'Don't play with words. This is an aeroplane.'

'Does your mother like the engine?' I asked him.

'My mother is an engineer. She must make it work.'

'Good,' I said, not grasping the real reason for his dissatisfaction.

'But I say to her you are no pilot so you need more power.'

Without an engine or the control wire to attach we occupied ourselves by lifting and shaking the Woodhopper to imitate the stress and load of flight. Apostoli remained moody and irritable. He disagreed with my suggestions and twice banged his thumb with the tools.

'Oh God what are you doing to me today?'

I proposed that we try to find the aeroplane's centre of gravity, by measuring one-third back from the wing's leading edge.

'This is an unscientific method,' he complained.

He wanted to suspend the flying machine from the ceiling, which was impractical.

'Is Katrin good at sewing?' he asked.

'It's not her strongest point.'

'Then you will have to do a lot of it.'

His promised four-point aviator's seat belt turned out to be an odd collection of parts from several automobiles which had been rusting in his grandfather's *apotheki*. 'American Buckle Corp. Jacksonville, Fla. Model 6975.' I tried to dismantle the assembly but the springs and ratchets were seized. Apostoli picked up the unit and threw it on the concrete floor, smashing its plastic casing. Then he prised off the locking mechanism, bending my favourite screwdriver.

We were not working well together. I offered to buy him a frappé.

At the *kafeneion* he played with his lighter, slipping it in and out of its silver sleeve. Then he muttered, 'I don't have a girlfriend any more.'

'What happened?'

'She says I am too boring for her. *Bah*, it is no matter. It is good to be single.'

The coffee did little to improve his mood.

'All weekend I am thinking about the control stick,' he said when we returned to the garage. 'It is a crazy design.' He mounted a pulley in the vice and pulled a length of string across it at a sharp angle. The string slipped out of the pulley. 'If the wire comes out then you cannot control the aircraft and you crash.'

'The original design worked.'

'The American way is to follow the plans, no nearer, no further. I will take the stick home and make a better control tonight.'

'Like the air-speed indicator?' I asked, making a point. He had promised to build that at home too.

'I will calibrate it tomorrow,' he said. 'Do not worry.'

I reminded him of the Woodhopper's performance statistics: take-off speed 27 mph, cruising speed 35 mph, stall 20 mph.

'I know,' he snorted, 'your Woodhopper will fly like a donkey.'

'Like a dragonfly.'

He kicked at the aeroplane's axle. 'I don't know why you did not refuse this stupid wood chopper motor.'

It wasn't a day for co-operation so, instead of designing a safer pulley system, we took each other's photograph sitting in the cockpit. He slipped on his dark glasses.

'I would be happier being a monk like Christodoulakis,' he sighed. 'Living in the mountains with a few goats.'

'Wouldn't you miss your friends?' I asked.

'Never. I would come down every three months and buy only things most needed. And maybe see a movie. Life would be simple and I would be happy.'

He looked at his watch.

'Now I go to the gym,' he said. 'My old girlfriend is being very sympathetic. I meet her and come back tomorrow.'

'With the engine?'

'Of course.'

But Apostoli didn't turn up the next day. Or that week. Not because he had lost interest or romance had blossomed but because the village's generosity had created new obligations of Byzantine complexity.

The gift of the engine had put Ariadne and Apostoli in the

villagers' debt. Ariadne was duty bound to make their Minirelli work, even though it was underpowered. If she failed she would lose face. Which was why, in her frustration, she broke the propeller, leaving her beholden to its owner, whoever he was, as well as the villagers. 'Are we sailing straight or is the shore crooked?'

It wasn't until the following Monday that Ariadne turned up in Anissari. She had been delayed by 'other responsibilities', such as drinking coffee and keeping the 340th Squadron airborne.

'You need to see the work I have done,' she said. She too was in a temper. Maybe it was the weather. 'I have lost sleep and today had no siesta.'

The village's little Italian engine was mounted on rubber mounts on a metal frame. A welded steel bracket had been made to connect it to the front of the boom. A new propeller had been joined to the engine by a spindle and bearing. The work was simple, clean and well executed. She had also rigged up a spring-loaded throttle.

'It's beautiful,' I said.

Ariadne said nothing, lit a cigarette then set the engine on the floor. Apostoli pulled the starting cord and it kicked into life, sounding insignificant.

'It makes smoke because it is new,' she said, raising her voice to be heard above the insistent buzz. 'I put a lot of oil into the fuel.'

Ash from her cigarette fell on the petrol tank.

'I ran it on the bench and calculated the rpm. It's about three horsepower.'

'Isn't that too little thrust?'

'Maybe you won't take off,' she said. 'Maybe you just go up and down the runway, though this is not for sure.'

The cultural inability to say no was at odds with my attempt to achieve a specific objective within a set period of time. The Minirelli engine was simply what had been available to the villagers at the time.

'How can I tell Yióryio?' I asked her.

'You cannot tell the village without causing offence,' said Ariadne. 'Which is why I am doing this work.'

'On an engine that isn't suitable.'

'There is no choice. On Crete we have motors for tractors and pumps but nothing like you need.'

At last a straight answer.

'In Greece we say if you want to build a house, build three: sell the first, give the second for rent and live in the third. This is how you learn. So your third aeroplane will be the best.'

27. On Top of Crete

Rafts of cloud drifted above the foothills of the Psiloritis. Katrin and I drove up into them, out of the sunlight and bird song, over a cool, grassy pass reminiscent of the Jura, towards the top of Crete.

In the Amari valley, the fruit bowl of Crete, hillsides were flecked with pink blossoms. Narrow, whitewashed lanes doubled back like Greek conversation. Here, wrote George Psychoundakis in his war diary *The Cretan Runner*, 'there was good wine, good company, sweet grapes and the gramophone playing in the shade of the cherry-trees'.

After the German invasion in summer 1941 many Cretans who had fought alongside the Allies took to the mountains. The English archaeologist and classicist John Pendlebury, who had succeeded Arthur Evans at Knossos, used his knowledge of the island to help organize a resistance network, arranging supply lines, storing sabotage equipment and seeking out coves for smuggling arms and men. About a dozen British intelligence officers, including the author Patrick Leigh Fermor, lived out the war in caves and *mitáta* shepherds' huts, liaising with and coordinating the efforts of the defiant Cretans. In Vaphé, a village near Anissari, one house became known as the British Consulate as so many stranded soldiers and agents received food and shelter there. But for the most part the resistance operated higher in the mountains, from where, in one audacious guerrilla strike, Leigh Fermor, Stanley Moss and half a dozen Cretans kidnapped the German commander General Kreipe.

In 1944 the Amari's villages suffered a terrible fate. In reprisal for an independent Cretan attack the Nazis razed every building on the western side of the valley. German troops went from village to village looting, shooting, driving the barefoot women into the hills and burning houses. Schools, churches and wells were dynamited. In Gerakari fifty-two men were executed across the dusty lane from the Sweetly Cherrys shop. In Ano Meros sappers blew up the cemetery. George Psychoundakis, watching from the cave above Níthavris, wrote that the burning went on for a week.

'They launched this cruel campaign to terrorize the entire island, and to show us that the Germans in Crete still had the power to destroy and overthrow, as barbarously as ever, all that still remained standing.'

The valley's husbands, brothers and sons were killed in a savage effort to break the Cretans' spirit. Less than a year later the defeated Germans withdrew to within a heavily fortified perimeter around Hania, willing only to surrender to British forces, fearful now of Cretan revenge.

Beyond beautiful, despairing Amari the Old National Road tucked itself into the foothills. Tumbledown terraces spilt down the valley, their stony edges unmaintained, eroded by centuries of weather. We climbed above the olive and cherry trees into an inland world of oaks and rock, up the fissured and wrinkled hide of the Psiloritis.

In Perama old men sat alongside wind-blown burial grounds. Nut-skinned gipsies sold shoes and tomatoes off the backs of decrepit vans. Above Pasalites serrated cliff tops rose like cockscombs, their steep flanks scattered with the scorched white skeletons of burnt trees. The charred torsos had been caught in mid-stride, as if trying to outrun the arsonist's flames. Near Zoniana, where locals were said to thrive on stealing sheep, a bloody billy goat head was impaled on a fence post. The road signs were peppered with gunshot. Katrin locked the car's doors.

Anogia was a leafy, highland town of widows and shepherds, part lost in cloud, part caught in humid sunshine. It too had been destroyed in 1944 and its surviving women kept body and soul together by selling embroidery. We drove through it, climbing still, the road winding across a vast, baked landscape of *mille-feuille* rock. There were no more settlements now, only the occasional low stone *mitáta* and unlicensed high-wheeled pick-up. The trees grew smaller, the air cooled by ten degrees and the earth appeared bereft of soil. Our ears popped with the altitude as we inched nearer to the sky.

Then, at 1,400 metres, we tipped over the lip of a hill. A wave of starlings surged up, circled above us and swirled back to earth. Shafts of sunlight chased manes of cloud through a gap between the ring of peaks, trailing vapour across the breach. Choughs crackled in the cliffs. A pair of eagles glided overhead.

The Nida plateau opened below us, a lush green sea lapping on the shores of towering, ash-grey mountains; timeless, spellbound, enchanted by isolation, by silence, by the sheer power of the encircling heights. It had been a sacred site since

Minoan times. Zeus was said to have been raised, even born, in its Idean cave. Pythagoras had visited it. Plato set *The Laws* along its pilgrimage route. Nikos Kazantzakis claimed to have lost his virginity on its summit.

On the new black tarmac lay great boulders as if the mountains, shivering at night, had shaken off a dead stone skin. We circled their rocky rim and dropped down onto the flat, sweeping plain. The treeless flanks were dotted with sheep, their bells ringing in the cool air. A solitary shepherd chased them on foot, his arms spread wing-wide, gathering the flocks until a wave of animals washed towards a waterhole.

I squatted down and touched the ground. It was hard. The grass had been shorn by grazing. There were some dry meltwater gullies and outcrops of sharp stones but for the most part the mile-wide plateau was unbroken and open.

'It could work,' I said to Katrin.

Except that the wild sky whistled around us, curling and rolling as if to the music of the air. I listened to it and remembered that early open-cockpit biplanes were flown to an extent by the sound of the wind in their bracing wires. It was said that whenever a student pilot slowed into a dangerous glide the wires hummed a descending melody of 'Nearer My God to Thee'.

'We'll put up a windsock,' I proposed.

'The wind is blowing from every direction,' said Katrin.

I wished she hadn't noticed.

'There must be moments of calm.'

'The earth is rough,' she went on, more realistic than I was about Nida's aeronautical potential. 'And it's at least a ninety-minute drive to the nearest hospital.'

Around us eddied fickle gusts, swinging from the south then from the west, even though there wasn't a breath of breeze anywhere else on the island. Flying near mountains had killed many unwary aviators, brought to earth by sudden downdraughts. Nida may have been the most romantic place to fly in Crete. But it wasn't perfect. Only, by now, I had no other choice.

'You like my mountains?' asked Kosmas, the weathered, crook-waving shepherd. He appeared before us, a pint-sized silhouette wading out of the sharp light. 'Sometimes I don't understand how they can be so beautiful.'

Since time immemorial men had driven their flocks to summer pastures here and to a solitary life in the mountains, after wintering on the milder lowlands. They had made rounds of sweet *mizíthra* over open fires and taken potshots at passing Turks. Their quick wit had earned them a reputation as Zeus's teachers, or so they claimed. Shepherds were revered in modern Cretan mythology too; independent, self-reliant and – nowadays – dressed in designer black. Most of them tended to be taller than Kosmas, striding into the hills with their *voúrias* backpack and Beretta pistol, but limited stature had, if anything, increased his cocky self-confidence.

'You have come to me to learn about the sons of the gods and sheep thieves,' he pronounced, drawing himself up to his full height. He ran stubby, dark fingers through his uncombed hair.

'No,' I said.

The few tourists who made the long journey to the plateau usually came to hear daring stories of the shepherds' life. Kosmas was prepared to oblige them for a small fee.

'Then you wish to climb Mount Ida and need a guide.'

'We're looking for somewhere to fly,' I said.

'No problem,' smiled the Lilliputian, unperturbed, revealing a set of tiny, pitted teeth. 'Zeus liked Nida – nice cave, good company, lots to eat – and he flew. So you can too.'

The alacrity with which my far-fetched intentions were accepted never ceased to amaze me. The Cretans, who had no inclination for moderation, had always believed that excess was divine and so condoned extreme behaviour. Which went some way to explaining why the Civil Aviation Authority cut no ice in the ancient Psiloritis.

'But won't the sheep get in his way?' worried Katrin.

'They will run,' Kosmas said, waving his hands to show their scattering. A stubble of beard darkened his pygmy face. 'Sheep run but men fly,' he laughed. 'And from the air you will see our Icarus.'

Kosmas strutted ahead across the green sea, a black-clad midget who seemed to encompass all that was raw and mortal in the mystical landscape. Katrin and I chased after him, over knot grass and the odd purple crocus not yet nibbled by the sheep. At the plateau's northern bank the grass licked against organized lines of boulders. Kosmas led us around them and onto a steep rise.

'You see better from here,' he said, bounding goat-like up the hill. 'But best from the sky.'

As we climbed the uneven stones fell into a pattern, taking on the shape of a plumed animal. Higher still the paws became heels and hands. The rock sculpture revealed itself as Icarus, or an angel, with body ascending 100 feet across the flat ground. Stabs of vaporous sunshine glided across its contours, from head to tail, and the rise and fall of shadows seemed to animate its stone wings. A gorse bush rustled over its head like hair blown back by the wind. The figure appeared about to take off.

Over the last twenty years a German artist, Karin Raeck, had created the impressionistic design to commemorate Cretan suffering at the hands of her countrymen in 1944. Her gesture of reconciliation, whether depicting a mythological boy, a winged Christian attendant or a flying freedom fighter, underlined Nida's ambiguous quality as a place between heaven and earth, myth and fact.

'You stay tonight,' said the dwarf.

The Taverna Nida was a monstrosity, a brash and invasive bunker more suited to suburban Moscow than a sacred mountain. It was the only man-made object on the plateau, apart from the sculpture, the few *mitátos* and the single-track road.

'In summer I work here,' he said, flapping his diminutive arms towards the broad plain. 'In winter I philosophize.'

He didn't invite us to supper. He simply moved a table to the cracked dining room window, produced a dozen hacked chicken limbs and sat us down.

'I tell you the truth about shepherding. God's truth. First, I have ten thousand sheep,' he declared, an exaggeration which we didn't dispute. His only sheepdog, a needle-toothed puppy, would have been hard pressed rounding up a mouse.

'Second, it is a hard life and we depend on no man.'

This claim was perhaps less of a fiction, if their exemption from income tax and reliance on European agricultural grants were overlooked.

'Third, we do not steal sheep.'

'We have been told otherwise,' I said.

'Lies,' said Kosmas. 'All lies. Unless the other man steals first.'

The ballroom-size cavern was bereft of any comfort, without cushions or armchairs. Along the back wall were three vast wine barrels, three video-game consoles and three *kri kri*, or ibex heads, one wrapped in tinfoil as if a take-away meal. Clutches of framed photographs of armed shepherds striking defiant poses surrounded the bar. At the table in a silent corner sat Kosmas' mother, grandfather and two tall sons.

In summers past shepherds' families had stayed in the home

village, the women and children riding into the hills only at shearing time. But with the advent of roads most shepherds put aside self-reliant tradition, if not the illusion of it, driving home every night to drop off the day's milk and watch television.

'My father was a shepherd and my grandfather before him,' Kosmas said with a gesture to the family group. 'My sons will be shepherds after me.'

The old man, fine-boned with a head of silver hair, played cards with his daughter-in-law. The elder boy, aged about fifteen, punched holes in a strip of leather and stitched a bell to it to make a sheep's collar. His brother showed more interest in shuffling through a stack of video cassettes. No words were exchanged between them.

'You will want to know about the *fourókattos*,' Kosmas volunteered.

I'd never heard of the *fourókattos*.

'The wild cat is famous among shepherds but no outsider believed in its existence.'

Wild cats had long inhabited Greek myths. Galenthias turned into a cat to become a priestess of Hecate, granddaughter of a Titan. The Psiloritis shepherds maintained that such an animal had always stalked the high mountains, taking their lambs at night. But the story, like those of Troy before Schliemann and Knossos before Evans, was consigned to the realms of fantasy. No other European had seen the beast, though two large, tawny skins were bought in Hania by the wife of a British scientist in 1905. Until 1996 when a team of Italian zoologists studying mountain carnivores snared a twelve-pound tiger-like cat.

'I saw it myself,' said Kosmas, anxious to claim his part in any heroism. 'My grandfather often shot at them from the *mitáta*.'

The discovery confounded the island's accepted zoological history, and proved the truth behind the shepherds' story. The feral cat did not belong to any European species. Its nearest relative inhabited North Africa. Either it had been imported as a domesticated giant by the Minoans or it had lived its reclusive life over millennia since the island had become separated from the mainland.

'No one believed us,' said Kosmas, 'and it was true.'

Kosmas' mother brought us cheese and olives, her movements quick like a little hen, then she sat back down to read a drug instruction leaflet. His grandfather unfurled himself from the table, picked up his crook and wandered out into the gathering dusk to tend to the animals.

'You say you are writing a book?' he asked, watching his younger son loading the video. 'Not making a movie?'

'That's right.'

The opening scene of *Zorba the Greek* had begun playing on the screen.

'I ask you because if the maker of that movie ever comes back to Crete I will slit his throat.'

'Why?' I asked, thinking perhaps that he couldn't forgive the film's contribution to the influx of tourists and erosion of Cretan traditions.

'Because it says that Cretans are vengeful and carry guns.'

'But they do. You do.'

'Maybe,' he shrugged, 'but that is our business.'

Kosmas dropped down from the table and pushed a handful of dry gorse into the fireplace. He lit a match and the wood crackled into flame. In spite of the day's heat it had turned cold in the voluminous bunker. In the fading light the last white snowfields on the black mountain looked like alien continents on an unfamiliar world map.

Kosmas and his silent family felt to me like refugees, under siege, driven out of their country by invasions. The Turks and Nazis may have been expelled but in summer some Cretans still retreated to mountain hideaways, or to behind reception desks. I imagined that Kosmas' wife – whom he did not mention – worked in a hotel on the coast.

'I don't take my sheep any more to the sea during the *chimadio,*' Kosmas said, refilling our glasses, 'because last time I killed a tourist there.' He smiled, gratified by our reaction. 'One time I saw a foreigner drowning in the sea,' he explained. 'I jumped into the waves and pulled him to shore. But he was so weak and hopeless that I pulled out my .38 and shot him.' His family didn't look up when he laughed. 'It was a mercy killing. I hate to see a creature suffer.'

I learnt later that the story had been invented by a local schoolboy, picked up by a journalist from Heraklion and reproduced in the Athenian papers. The child's personal fantasy had caught the public imagination and grown into a popular myth, which every shepherd then adopted as his own.

'But I don't need to shoot you because you will kill yourself in your aeroplane,' said Kosmas in an attempt to reassure us, 'and we will all tell the story.' As he fed his chicken bones to the puppy he asked, 'When is it you want to fly?'

'Next weekend?' I said.

An hour later the electricity 'went on holiday'. Kosmas led us through the dark, concrete warren by candlelight. 'You like the room?' he asked, the iron bed and tourist board prints jumping in the flame.

'It's fine,' I lied. 'Thank you.'

The cold, dank cell could not have been further from the open beauty of the plain. The sheets were damp and not very clean. I imagined the mountains shivering in the night and a rock crushing us in our bed. In the next room a lamb cried to be fed.

In the morning we drove home to Anissari to prepare to fly in the sullied preserve of the gods.

28. Clear Prop!

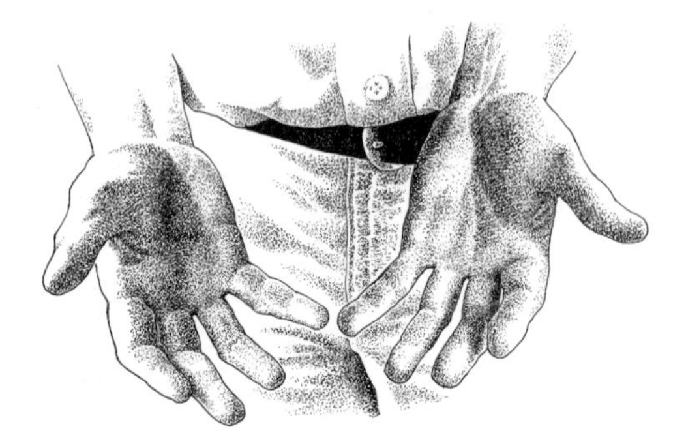

On Monday I flew to England to borrow an engine.

I found it in an hour. John Hamer, a retired RAF test pilot and microlight importer mentioned in *Flight*, confirmed that the Minirelli wouldn't work – unless I flew at 400 mph and weighed twenty kilos. The Woodhopper would have sat on the runway, its engine buzzing, not moving forward, with bits falling off.

'Do you have anything suitable?' I asked him on the phone.

'I don't know you from Adam,' said Hamer, 'and if you crash on the airfield the Greek aviation authorities will seize my gear.'

Airfield? Authorities?

'I suggest you find a paramotor.'

Paramotoring was the sport of strapping onto one's back a kite-like wing and caged propeller and stepping off a high cliff. It made microlights seem suburban and safe. Their engines were lightweight, portable and suited to the Wood-hopper.

I called Par Avion who suggested Perfect Bore who put me in touch with Bailey Aviation in Royston. Paul Bailey, a former national karting champion and open-air flying addict, asked the weight of the Woodhopper, its take-off and cruise speeds. I heard his pencil scratching on the back of an envelope. 'You need about fifty kilos of thrust,' he told me. 'Our JPX D320 – with a 1.09-metre prop – produces sixty-four kilos.'

'Great,' I said.

'The only problem is that there are eight people on the waiting list in front of you. And I think the exhaust will blow right in your face.'

'I need an engine by tomorrow,' I said.

'I do have an old Solo 210 with two hundred hours on it,' he volunteered. 'But I don't know if I have a prop. If I do, I'll lend them both to you.'

Ten minutes later Paul found a three-blade carbon Kevlar propeller. I called Ariadne to ask her about its suitability. 'Sorry I didn't answer when you rang first time,' she said. I heard the sound of lapping waves and laughter. 'I was swimming.'

On Tuesday morning I drove to Hertfordshire. Paul ran the engine on his test rig. It whirred like a turbo-charged sewing machine. I loved it, and it scared me to death. As he dismantled it for shipping he wrote 'front' on the face of the prop.

'Take care of yourself,' he said. 'The engine's replaceable.'

I drove to Heathrow, packed the engine in an old suitcase and checked it through to Hania. I carried the prop as hand luggage.

'I've seen a lot of nervous passengers in my life,' said the security guard, 'but you're the first carrying his own propeller.'

The suitcase wasn't stolen in Athens or inspected in Hania. Katrin collected me at the airport and we drove to Anissari, sneaking the engine into the garage without any villager suspecting that their feeble motor wouldn't power a hairdryer. It too was a very Greek solution.

On Thursday I made the control system. In the half-baked plans the cables seemed to slice through the tail's horizontal wing. The resulting holes would have destroyed the aeroplane's lift. With a length of string I worked out an alternative route, rigging pulleys on the aluminium down-tube and running cables back to the rudder and elevator. Apostoli arrived as I was attaching the control stick. My control stick.

'Good morning,' he said. It was mid-afternoon. 'What is happening?'

I remembered that living in Greece required a tactical mind.

'You have worked so hard on the Woodhopper,' I told him, 'that I borrowed this engine to make us successful.'

My gambit burnished his pride. Spurred into action he rummaged in the bottom of my parts box for angled brackets, plates and pulleys. Like a child playing with toy blocks, he invented an ingenious junction for the elevator cables. Its most unexpected quality was that it worked.

We set and tensioned the controls in three hours, ensuring that rudder and elevator were straight with the stick upright. I pushed us forward, telling him that we had to finish the control system that day. And we did.

Then we mounted the new engine on the Woodhopper, stowing the villagers' motor out of sight.

'It looks like the first aeroplane,' Apostoli said.

'I want to fly it at Nida on Saturday,' I told him.

'Nida?' His excitement for the engine was as great as his horror at the idea of my lifting off from the plateau. 'Nida is not an airfield. It is a mountain.'

'It's my only option,' I said. 'I'm out of time.'

'I have an airfield for you,' said Ariadne.

The sudden concertina of events exhilarated me. I'd begun Friday morning by making my final To Do list: 1. Mount wheels. 2. Design air-speed indicator. 3. Buy foam for seat (don't use Yióryio's cushions). Ariadne and Apostoli arrived before ten o'clock.

'You mean the Nida plateau?' I asked her.

'Not Nida. Your airframe will not survive a take-off from any surface other than asphalt.'

'Where then?' I asked.

'I cannot tell you.'

'Then how do I get there?'

'You follow the truck.'

'What truck?'

'The one you must find tonight.'

I asked her if the Ministry of Defence had granted us permission.

'We don't need permission because your flight is illegal,' she replied.

'I don't understand.'

'If nobody knows about it then we can't ask for their approval. The only request is that when you fly your circuit, you turn over the sea so you are not seen.'

'How can I take off from an active airfield without being seen?' I asked.

'Because it is the weekend and everyone will be at the beach. It is the best way.'

'Be happy,' said Apostoli. 'We have an airfield.'

I didn't know how it had happened. Maybe a base commander had relented. Maybe Ariadne feared losing face. Or maybe she understood that now I would fly with or without her help and this was her way of preventing me from breaking my neck. But, whatever the explanation, I had a runway, somewhere, and that was better than a mountain.

Apostoli set to work designing a simple air-speed indicator: a sprung pendulum scoop which would be pushed back by the airflow.

'Now I fix it to the side of a car and drive,' he explained.

By lunchtime he had calibrated it, avoided hitting a flock of sheep and sworn every drinker in the *kafeneion* to secrecy.

'No one must know where the Woodhopper flies,' he told them as they bought him glasses of *tsikoudiá*.

'Apart from the television crew,' Yióryio reminded him.

'What colour helmet do you want to wear tomorrow?' he asked me. 'I have four different ones.'

'I'll wear my bicycle helmet,' I said.

'Does it have an oxygen mask?' he asked.

Yióryio took me to meet Athanásis. The village truck, which stank of goat and usually carried irrigation hoses, would help to launch the maiden flight. We settled on a price, as everyone in the village except me knew where I was heading, and Athanásis agreed to be at the garage by six the next morning.

At home Katrin grilled swordfish, marinated in olive oil and lemon, and served it with parsley and tomato salsa, lentils and *hórta*. We didn't speak. After supper I walked alone through our adopted valley, listening to the summer's cicadas, beneath the darkening mountains. My feet led me to the garage and my light, white flying machine.

I sat in its wooden seat and twisted the stick left and right, up and down, the rudder and elevator moving in response. I pictured myself making the small, fluid hand movements while flying forward, the elements rushing over the wings.

In *Wind, Sand and Stars* Saint-Exupéry wrote, '. . . once men are caught up in an event they have no more fear of it. Only the unknown terrifies men. Once confronted, it is no longer the unknown.'

I looked up. In the half-light Polystelios, who had hidden himself away in the weeks since Aphrodite's death, stood in the door frame.

'*Yiá sta khéria sas*,' he wished me. Health to your hands. 'May the sun shine and you not fly too close to it.'

29. Up, Up and . . .

The garage was smothering, hot and dark. The village was still asleep. I stood on the threshold staring into the shadows at the Woodhopper. Katrin and I had split it into three parts – the fuselage and two wings – ready to be moved to the airfield. Under the single light bulb I checked my tool bag for spanners, hammer, spare nuts and bolts. In a cardboard box were my bicycle helmet and decorator's goggles. I was ready to fly.

The inky sky began to glimmer above the Lefka Ori. The village truck, an old Hino which had been hand-painted turquoise, drove into the square with Athanásis behind the wheel. A cockerel crowed. A minute later Apostoli arrived in his silver Fiat. He looked like death.

'Are you all right?' I asked him.

'I haven't been to bed yet,' he said, not lifting his dark glasses.

Out of the door came the wings, carried above our heads like Christ's Easter bier. Our muscular Olympian, spent after a night's copulation, led the procession. Katrin, as numb as a sleepwalker, and I brought up the rear. To make more space Athanásis took a mallet to his truck and knocked off the spare wheel.

We laid the white aerofoils on the flatbed, devised a frame of offcuts and balanced the fuselage over the top of it. Against the dawning landscape and outside at last the Woodhopper looked small and insignificant, having filled the garage and our imaginations for so many months. Athanásis tied two plastic streamers onto its tail.

'In pieces it looks ugly,' carped Apostoli from behind his Ray-Bans. 'Like a fledgling bird.'

As he lashed down its parts, I turned back to the garage for the last time. It was an empty space again. In the far corner stood two tins of emulsion, an orange paint tray and the tail end of a roll of Ceconite, covered with cobwebs. My footsteps echoed in the stuffy cavern. Flakes of whitewash drifted down from the ceiling. I closed the heavy door and left it to the swallows.

'Follow me,' said Apostoli. Our casual loading had taken forty minutes. 'And tell no one where we are going.'

We began to inch forward in convoy, my car bringing up the rear, guiding the teetering load around the mulberry trees and under strings of fairy lights. The tick of the diesel echoed off the walls of the waking houses. Keys hung in door locks. Chryssoula opened her shutters. Ulysses glanced up from under his cowlick and babbled. Leftéri chased after us wearing my tinny Junior Pilot badge and holding my model Spitfire above his head.

Outside the *kafeneion* Yióryio waited, yawning, a home video camera under his arm. His television friends hadn't turned up but, despite all his insistence, he seemed not to mind. He alone would witness the flight for the village. The early drinkers helped him to load his ladder onto my roof rack.

'*S'tó kaló*,' Socrates said to me. Go with the good. Then with barely a backward glance, they dispatched us, before tending to the more important matters of picking fruit and sharing gossip.

Dawn, soft as powder, dusted the pale grey limestone of the White Mountains. Shaggy green cypress trees rose jet black into the bright, inspiring sky. The Woodhopper's white delta tail flipped in the breeze. Athanásis skirted a line of trees and my flying machine was all but whisked off the flatbed by a whipping branch.

At Vrysses we turned west onto the National Road. Apostoli accelerated to 30 mph but the aeroplane's wires shivered with such violence that I flashed my lights to make him reduce speed. At a snail's pace we skirted ancient Aptera, where the Sirens had lost their wings, and modern Akrotiri, with its four-mile-long Space Shuttle runway. At the Privilege nightclub Apostoli slowed to wave to clubbers with whom he'd been dancing a few hours earlier. As the sun rose at our backs we followed the coast road, along the sea into which Icarus had plunged. The day promised to be a scorcher.

Ariadne was waiting beneath pink-crepe bougainvillaea at the perimeter gate, which swung open to admit us. Armed guards in combat fatigues directed us between the ranks of fighter aircraft: T-33s, Starfighters and a Mirage F1C Interceptor. The rustic truck followed Apostoli past the barracks, across the apron and onto the runway. We turned along it, drove 100 yards and stopped.

'This is our place,' said Ariadne. 'For today only.'

There was no hangar, power point or toilet. Thyme and sage lined the long, deserted airstrip. A single windsock drooped above the rocky shore. Martial music echoed in the distance. It was eight o'clock and baking hot.

'Let's cut the mud,' said Ariadne. Meaning hurry. 'It will be more than thirty-five degrees today.'

With no spoken plan and few words we set to work, unloading the truck, disentangling the wires and unwrapping the wings. Each was lifted in turn and attached to the frame. Apostoli and Yióryio bolted on the self-locking nuts, the clinking of their spanners resonating along the drum-like length of the wings. Clips and washers rang as they dropped onto the tarmac and the shrill of cicadas rose with the gathering heat.

'I am in paradise now,' Apostoli sighed, breaking our silence. 'Today I will see the Woodhopper fly.'

'Nice to hear,' said Ariadne to her son, 'but if it crashes he will make *souvláki* of us all.'

As Apostoli tightened the bracket bolts I lifted the engine out of its suitcase and passed it up the ladder to Yióryio.

'Other way around,' instructed Ariadne. 'Come oooon.'

We bolted on its three-blade propeller and attached the throttle. I ran the fuel line along the boom to a one-litre plastic bottle taped to the back of the seat. Katrin connected the stop switch above my head. In two hours the flying machine was reassembled, my Cretan family working together. Under the high, azure sky the Woodhopper had regained its beauty. The sun shone through the fabric wings of our simple white dragonfly, quivering beside the sea.

'I forgot to buy petrol,' I said.

'Drive up to the gas station and get it,' suggested Ariadne, 'if you haven't anything else to do.'

At the gate the guards let me out and then back into the base. In the shade under the wings I mixed the fuel with two-stroke oil. I poured the liquid into the litre bottle, spilling a cupful on the tarmac. It evaporated in seconds. I primed the engine, a plume of petrol squirting through the carburettor and missing Yióryio's cigarette.

'Why don't we give it a try?' suggested Ariadne.

I sat in the pilot's seat wearing shorts. In her flip-flops Katrin put her foot against a wheel. Apostoli held the tail and Ariadne moved away the ladder. 'Contact,' she instructed with a nod to Yióryio. When he looked blank she said, 'Go.'

Yióryio yanked at the starter cord, shaking the Woodhopper from nose to tail. I nearly fell out of my seat.

'*Ela moró mou*,' said Yióryio. Come on, baby.

'Less throttle,' said Ariadne.

I clicked on the seat belt.

He pulled again. And again. The Solo 210 refused to be started in a hurry, especially with the fuel evaporating in the heat. Ariadne checked the spark plug and adjusted its mixture. Apostoli asked her about the vacuum in the crankshaft. The two-stroke needed to be wooed, cajoled, persuaded, until we were all dripping with sweat.

'*Ela koúkla mou*,' said Yióryio and pulled one more time.

The engine leapt into life, sounding like a bag of nails. The frame shivered as if an electrical charge had shot along its length. The wash from the propeller pushed me back in the seat. I tightened the seat belt. As the engine warmed, the wood absorbed the vibrations. I checked the free movement of the controls. I felt both exposed and at ease.

'Slow taxis only,' instructed Ariadne.

'You have only fifteen minutes' fuel,' yelled Katrin over the din.

Apostoli let go of the tail. I lifted my feet off the tarmac and balanced them on the axle. My hand-built aeroplane began to creep forward under its own power. I whooped with delight. Katrin began to cry, awash with emotion, watching our creation carry me away.

I twisted the stick, turning the rudder and curving into the middle of the runway. To the left was the Cretan Sea. Ahead of me the Lefka Ori were a hazy silhouette against a molten midday sky. The broad, white, luscious centre line slipped beneath my feet. There was no sea breeze, thank goodness, for the Woodhopper was too much like a glider and I was not enough of a pilot.

I increased the throttle and picked up speed, feeling every bump and blemish on the tarmac. The tail wheel, which was a castor from Yióryio's sofa, ran over a small stone and shook all the wires. I was not going to fly. Not yet. The intention was for me to get a feel for the handling, for speed, for turns. It was difficult to travel in a straight line.

Near the end of the airfield I reduced the revs but the aircraft didn't slow down. There were no brakes. Santos-Dumont had worn reinforced gloves to grab his tyres to stop his Demoiselle. I touched my heels to the ground and brought my Woodhopper around, leaning into the turn as if surfing on a turbocharged shopping cart.

I racked up the throttle and began to sail along the length of the airfield. The aeroplane quickly gathered speed, sweeping me forward. The speedometer fell off. I thought of the early flyers who didn't understand the physics of flight. Much like me. Had they been moved by the same sense of horrified wonder? The propeller blades turned at 600 mph less than three feet in front of my face. The stick had to be forced to the right. I managed to keep to the centre line.

But I felt no sensation of lift, which concerned me. I brought her round at the far end of the field and idled the engine. Katrin and the others watched from the side of the runway. I wanted to do a faster run. I needed to know if the wings' shallow profile was sufficient to provide lift. I couldn't wait. I pushed her to three-quarters power.

The Solo's tone changed from angry sewing machine to lively lawnmower. I eased the stick forward to cancel any lift, to keep the wheels on the ground. In a few seconds I passed the crew at maybe 20 mph. I stretched myself to feel the lightness, imagining the wings as extensions of my arms. I wanted to sense the first hint of uplift, to begin to leave behind my heaviness. I was moving fast. The air rushed past me.

Then a tyre blew out. Its hub snapped. The speed fell away. I swung the Woodhopper around and limped back to the group. I cut the engine and everyone spoke at once.

'I didn't feel any lift,' I shouted, my ears ringing.

'The wheel's smashed,' said Katrin.

'Where's my speedometer?' asked Apostoli.

Yióryio didn't think I'd reached take-off speed – 27 mph.

Ariadne said, 'You will have to do short hops to find how the aircraft flies.'

'*If* the aircraft flies,' I said. 'The engine may not be powerful enough.'

'It needs a new spring,' said Apostoli, having retrieved his speedo. 'And I need my bed.'

We had worked six hours without a break. The air temperature was spiralling. My arms were sunburnt and Apostoli was asleep on his feet. Ariadne advised that we return to Hania, have a siesta then try again in a couple of hours.

'But we need new wheels,' said Katrin anxiously.

'Don't worry, I have the perfect ones at home,' she said. 'Exactly the right size.'

The dense heat thickened my thinking. There was no shade at the airfield. Ariadne suggested that we reconvened at six o'clock. I proposed five o'clock, knowing that the Greeks would be at least an hour late. We would then have three hours of daylight to fit her wheels and to fly.

We left the Woodhopper under military guard and the metal-hot sky. In convoy we drove out of the gate and headed towards town. Katrin and I opted to go swimming and peeled away from the others, falling into the sea and rinsing ourselves awake.

As we dried our bodies in the sun I said, 'I want to see those wheels.'

We hurried after Ariadne, reaching her house before she went to bed. Apostoli's snores resonated from behind his bedroom door. In her *apotheki* under a dozen nubile, dancing Nereid dolls were the wheels. Their ballbearing bore measured two centimetres in diameter. My axle was three centimetres thick.

'I was thinking that they were too small,' she yawned. 'But you can drill them out.'

The wheels were steel. There was no electricity at the airfield. I looked at my watch. It was almost two o'clock. In thirty minutes the shops would shut. If I didn't fly today another week would pass, the winds could start and I might go doolally.

Katrin and I hurried first to Knossos Motor Parts, which had closed early for a christening, then on to Sidal Supply, a tool and hardware store. On their racks were dozens of tyre and wheel combinations: tubeless, barrow, ATV and trailer. None of them was the right size. So I bought two high-speed, pneumatic assemblies with an overlarge four-centimetre bore. I'd hit on the idea of slipping an inner sleeve between my axle and the bearings. Of course, Sidal didn't sell tubing. So I grabbed a hacksaw and rat-tail file in case the idea didn't work.

As shopkeepers closed for the weekend we raced towards the central market. I now owned nine wheels: Ariadne's pair, the two from Sidal, plus various Anissari wheelbarrow and *mechanáki* models, broken and unbroken. Mariada Papadaki was Hania's oldest 'industrial shop'. Its constancy might make it the last to close. We pulled up outside it as the shutters were coming down.

'I know where a hundred and fifty thousand different items are in my store,' said Papadaki, after submitting to our pleas and admitting us to the darkened emporium, 'so why can I never find my spectacles?'

I bought half a dozen lengths of copper tubing, an air pump, a jar of tumble-dryer grease and an extra trolley castor. Just in case. A goldfinch sang in its home-made cage. We paid, lunched at a street corner on an ice cream and returned to the airfield.

By the time the others arrived it was almost six thirty. The new wheels and their copper inner sleeves were attached, wires adjusted and the engine refuelled. The air temperature had dropped and there was no breath of evening wind.

'I say to my only God now you will fly,' declared Yióryio.

'Keep straight on the runway. Don't push your nose up. Don't put the nose down,' said Apostoli, rested and full of both himself and advice once again.

Two less helpful recollections sprang to mind. The first was from the Woodhopper's 'Learning to Fly' instructions. 'Make your first flight over a wheat field as the wheat will cushion any sudden fall.' I knew of no wheat fields in Crete. The second, from a motorcycle safety handbook, was that an unseated rider loses seven millimetres of bone for every metre of tarmac 'travelled' in an accident.

'I know you won't kill yourself,' Ariadne said as I changed out of my shorts. 'Before, it was a crazy idea, a utopia. Now it will be OK.'

'But nothing has changed,' I said.

'You have changed.'

I twisted myself between the wires and into the seat. Yióryio videoed the group 'so we count how many we are now and at end'. He put aside the camera and yanked the starter cord. The engine caught on the third pull.

'Just land,' said Katrin. 'Soft or hard, it doesn't matter, just land.'

There was nothing more to be said. I eased on the power and the Woodhopper began to move. The controls were responsive. The new wheels gave a smooth ride. Assuming that I had enough power, the flight would be a matter of coordinating eye and hand movement. I hoped that the wings wouldn't snap off.

I taxied to the end of the runway and brought around my flying machine. With my feet planted on the tarmac I revved the engine to burn off excess oil. My seat belt felt snug. The forty-eight wires looked sound and taut. I was the Red Baron in knee pads. Orville Wright in cycling helmet. Icarus himself wearing decorator's goggles. A son saying goodbye to his mother. I opened the throttle and lifted my feet off the ground.

The Woodhopper carried me forward with a surge of life. I laughed out loud at the sheer joy of it. I remembered the twisted tail and eased the stick to the right to compensate for imbalance. I had a sense of travelling at wild, almost reckless velocity. I remembered that flying was done largely in the imagination, the wings pushing down the air, the one element that can't be seen.

In less than fifty metres the machine began to feel lighter. The tyres seemed to be less and less in contact with the earth. The wing tips and tail started to rise. My heart lifted with them. Then the engine tone changed. I looked down and saw that the wheels had stopped moving. The air-speed indicator read 30 mph. My shadow swept across the tarmac beneath me.

I was flying.

Almost at once the right wing lifted sharply. The left wing

dragged behind it until I was flying at forty-five degrees to the ground. To compensate I pushed the stick further away from me. I threw my weight to the right. But the Woodhopper didn't rebalance itself. Instead, the angle became more acute. Motion was in my hands yet beyond my control. I was soaring sideways in a brittle aluminium shell.

Power controls height. Speed is governed by the elevators. I had no choice but to reduce power.

In seconds I came down on the new left wheel. Because of the acute angle its hub jammed. The bearings seized. The momentum pivoted the fuselage. The tail described an arc above my head. The nose pitched into the tarmac. The unbreakable Kevlar prop disintegrated, whipping its three blades away into the reeds. The aircraft skidded forward, the metal frame folding like a deck chair around me. With me inside it. At my back the weight of the boom pushed me towards the ground. I crumpled onto my knees, still strapped to the seat. I smelt petrol and thyme. In my ears I heard screaming. I reached up to switch off the engine.

Silence.

'I'm OK,' I shouted. 'I'm OK.'

I was alive.

Apostoli sprinted the 400 metres to reach me first. Yióryio followed him, yelling, 'How is your legs?'

I couldn't turn my head to see them.

'Fine.' My knees were half an inch off the runway. 'I'm fine.'

I felt whole. No limbs were missing. I heard Katrin's flip-flops slap across the tarmac.

'Get out,' ordered Ariadne.

'I'm trapped,' I chortled, in shock. Petrol hissed against the hot engine block.

'Get out now. We hold it up.'

As they lifted together I released the seat belt and crawled out from underneath the Woodhopper. My knees appeared to work. The tarmac felt hot and gritty against my palms. I stood shaking on my feet.

'I couldn't keep her level . . .' I told Ariadne.

'You did the best thing you have done,' she replied. 'To be in one part. You are in one working part.'

Katrin flung her arms around me, sobbing.

Now I could see the damage. The subframe was mangled and snapped. The tail pointed to the heavens like an arrow. My stick insect flying machine had come to earth, almost as if acting on my behalf. It had been built to fly and crash, to ground me. It was best to have done that at a low speed and from a low height.

Apostoli inspected the wreckage. 'Now we have the lesson to see that both wings need the same angle of attack,' he expounded, rather late in the day.

It transpired that the wings had been mounted at slightly different angles, making an inch or two's difference over the thirty-two-foot span, giving the right wing greater lift. The imbalance could have been corrected by ailerons. Except the Woodhopper didn't have them.

'Maybe next time,' roared the base commander who had appeared among us. Until the flight's completion he had been away 'fishing'.

'We need only a new propeller and some tubing to fix it,' said Apostoli, taking up the thought. He lifted a snapped length of aluminium and a shower of filings dropped to the ground. The wings, wires, tail and engine appeared to be sound. 'We try again next weekend?'

'No problem,' encouraged the commander. 'Everything is under control.'

It seemed fitting that the motto of his fighter wing – and the airfield – was 'Fly low hit hard.'

'I don't think so,' I told them, imagining myself impaled on a minced length of Greek tubing. 'Once is enough for me.'

'1–2-3–4-5,' Yióryio counted. 'We all here. I am happy.'

At the moment of the impact he had dropped the video camera to watch. In shock and relief Katrin had said, 'My love! Thank God!'

'You cannot believe how lucky you are,' said Ariadne.

I inhaled and the air seemed to sear my lungs. I felt as if I hadn't taken a deep breath for months. Yióryio uncorked a bottle and sprayed me with champagne: the first aviator to be fêted for crashing an aircraft.

'You know I do not drink alcohol,' said Apostoli, 'but I make an exception for champagne.'

30. Elements

I tucked my toes over the edge. Head up. Eyes open. Then sprang up and away from the wall, swinging my arms above my head. In that moment I was weightless, free of gravity. The sun warmed my back. There were voices in the distance. I fell towards the sparkling reflection, diving out of the air and into the sea, sloughing off a hide of silver bubbles like a snake losing its skin.

I had wanted to fly higher. My aim had been to make a series of hops and then to complete a single circuit of the airfield. I had wanted to touch the sky. In aeronautical terms the project was a failure, an expensive way to learn the laws of physics.

But the objective had never been to secure my future as an aviation engineer. I hadn't set out to become a pilot. I'd simply set my heart on lifting off the ground in an attempt to lighten my despair. I'd wanted to bid farewell to my mother and to recognize the power of love and life to transform sadness.

Then, with a ritual dip of my wings, I'd wanted to come back down to earth.

And I did that, nose first on the island where man first flew. I took to the air. I ascended, though not very high or for very long. Maybe I flew as far as had the Wright brothers' maiden flight, some 120 feet, or equalled the duration of the de Havilland Comet's first hop along the runway. Then again, maybe not. I hadn't carried a tape measure aloft. It was all over in a meanness of seconds. In the process I broke nothing that wasn't already broken, apart from odd bits of Kevlar and metal.

In the cool water I felt my tension shiver away. I swam forward underwater holding my breath, towards luminous patterns of sunlight and shadow, intimate and enormous.

In the few seconds of my flight, suspended between earth and sky, I'd half-hoped to hear my mother's voice. I'd imagined her flying alongside me saying, 'Well done, darling, but do keep your nose down.' I had wanted to tell her the story of the Woodhopper and our time in Anissari, its tall tales and hot, wine-soaked days. I'd hoped to say that I missed her. Of course, I'd heard nothing above the buzz of the engine for there was nothing to hear. No voice had called me from the silent dust. Instead my solitary flight, and the good fortune of my survival, encouraged me to a new life.

I broke the surface. White legs kicked at the water, treading down air and laughter, filling the sea with light and sound. I swam towards the voices.

A dozen children, and their guardians, splashed in the shallows at the edge of the bay.

'He flew as far as Icarus,' one child shouted to another.

'Icarus crashed into the sea,' said an adult.

'Did he die too?' asked a boy.

A young girl's long, black hair glistened in the sun. I couldn't imagine building and flying my aeroplane anywhere other than in elemental Crete, on the island where the Minoans had reached out to catch the essence of the passing moment, among a people with a raw, unpredictable, admirable energy for life.

The young swimmers surrounded me. A pair of boys were throwing a beachball. A breath of wind lifted it towards me. I caught it and, thinking of extending their game, lobbed it towards the black-haired girl. The ball landed in front of her. Laughing with surprise, she reached out and missed it. She stretched forward again, grabbing at the air. I realized then that she couldn't see the ball. The two boys, who were deaf, dived after it, squealing with delight. There was a wheelchair on the beach.

Beyond all the crippled children the sun caught the wave tops, covering the surface of the sea in ten thousand flashes of light; radiant, moving, alive with all that had come before, and will pass in time, sparkling, broken-hearted lives on a summer sea.

Postscript

Finn Cameron MacLean was born ten months and seventeen days after the Woodhopper's maiden flight.

Acknowledgements

I am indebted to the *filóxenoi* Cretans, descendants of the gods, who fed me, aided me and shared with me their remarkable stories. As another Greek noticed 2,500 years ago, fiction is woven into all. *Sas efharisto olous therma apo kardias.*

As for mortals, John Craxton, the British Consular Correspondent in Hania, provided invaluable assistance. Barry Blackmore advised me on the disadvantages of jumping off cliffs. Christine Gettins and Maureen and John Freely propelled me with early enthusiasm. Matthew Belfrage, W. L. Boynton, Malcolm Cockburn and Sam Kemp helped me to see DIY possibilities beyond cockeyed bookcases. Peter Kent shared with me the joy of jet-powered, exploding paper gliders. Ian Marshall carried me aloft aboard his *Flying Boats*. Antony Beevor, John Hamer, Paul Bailey and Guy Gratton, Chief Technical Officer of the BMAA, did their best to help me not to crash.

I had it easier than earlier Icaruses, my travel arrangements smoothed by Marina Yannakoudakis of ViaMare Travel, ANEK Lines, Aegean Airlines and Chris Laming of P+O Ferries.

The in-flight story would not have taken off without my agent Peter Straus, my radio producer Mary Price and friends David Chater, JoAnne Robertson and Joanna Prior. Juliet Annan and Mary Mount welcomed me to Penguin with soaring enthusiasm. I am grateful to Rachel Calder and Michael Fishwick for ten years' buoyant support. The Reverend John McMinn and Sue Botterill and Carol Austin of

CancerCare Dorset uplifted me with their sheer kindness. To them all, and to Piers and Silvia Le Marchant, thank you for entering wholeheartedly into the spirit of this madness.

I am also beholden to the Royal Literary Fund for their generous support at a time of need.

Only in Crete, the rugged, hospitable, achingly beautiful land of individuals, could this journey have taken wing. And only with the support and courage of my wife Katrin, who shaped the airfoils and attached the wires to the machine which might have killed me, could I have flown, and come back down to earth.

DISCLAIMER: I do not recommend or encourage anyone to build and fly their own aeroplane without first undertaking extensive training and preparation.

www.rorymaclean.com